FOURIER TRANSFORM INFRARED SPECTROSCOPY: DEVELOPMENTS, TECHNIQUES AND APPLICATIONS

CHEMICAL ENGINEERING METHODS AND TECHNOLOGY

Additional books in this series can be found on Nova's website under the Series tab.

Additional E-books in this series can be found on Nova's website under the E-books tab.

CHEMICAL ENGINEERING METHODS AND TECHNOLOGY

FOURIER TRANSFORM INFRARED SPECTROSCOPY: DEVELOPMENTS, TECHNIQUES AND APPLICATIONS

OLIVER J. REES

Nova Science Publishers, Inc.

New York

NOTICE TO THE READER

LIBRARY OF CONGRESS CATALOGING-IN-PUBLICATION DATA
Fourier transform infrared spectroscopy : developments, techniques, and applications / [edited by] Oliver J. Rees.
p. cm.
Includes index.
ISBN 978-1-61668-835-6 (hardcover)
1. Fourier transform infrared spectroscopy. I. Rees, Oliver J.
QD96.I5F685 2009
535.8'42--dc22
2010012203

Published by Nova Science Publishers, Inc. New York

CONTENTS

PREFACE

Fourier Transform Infrared Spectroscopy (FTIR) is a powerful tool for identifying types of chemical bonds in a molecule by producing an infrared absorption spectrum that is like a molecular "fingerprint". FTIR is most useful for identifying chemicals that are either organic or inorganic. It can be utilized to quantitate some components of an unknown mixture, as well as to the analysis of solids, liquids, and gasses. This book presents topical research in the field of FTIR including an overview of recent applications of FTIR spectroscopy in combination with chemometrics in the analysis of various quality parameters of fats and oils, a modified FTIR method for the analysis of various structural dynamic problems and energetic materials on surfaces.

Chapter 1 - Infrared (IR) spectroscopy is one of the vibrational spectroscopic techniques based on the interaction between electromagnetic radiation (EMR) and matters in the IR regions. It is rapid, sensitive, non destructive, relatively simple, ease of sample presentation, and can be easily adapted to be performed by untrained users in some laboratories. In recent years, because of the development of Fourier transform infrared (FTIR) spectroscopic instrumentation, the application of this technique has increased in the study of fats and oils. The main advantages of FTIR instruments over dispersive instruments are that they have increased sensitivity (high signal to noise ratio), permitted much higher energy throughput, and considerably improved the speed of spectral acquisition. The advance of FTIR spectroscopy has attracted some scientists for qualitative analysis of fats and oils owing to the large amount of structural information that can be extracted from the infrared spectra, and because of the "fingerprints" of functional groups that can be displayed narrowly and intensely in the mid IR region (4000–400 cm^{-1}). Combined with improvement of computer software and chemometrics, FTIR spectroscopy can easily manipulate the spectral information and become a powerful analytical technique for quantitative analysis. In this chapter, we describe an overview of recent applications of FTIR spectroscopy in combination with chemometrics for: (1) analysis of various quality parameters of fats and oils; (2) analysis of minor components and contaminats in fats and oils; (3) authentication of fats and oils; and (4) monitoring the stability of fats and oils. The basic principles of IR absorption and some techniques related to FTIR spectroscopy are also highlighted.

Chapter 2 - A modified Fourier series method is described for solving various structural dynamic problems. To better understand the essence of this new approach, a brief review is first given of the conventional Fourier series expansion and a few related mathematical theorems. An improved Fourier series representation is then present which can be used to expand any function, over a solution domain including the boundary points, with a pre-determined rate of convergence. Thus, for a given boundary value problem, an exact continuous solution can be systematically obtained by letting the series simultaneously satisfy both the governing differential equations and the boundary conditions on a point-wise basis. This improved Fourier series method is first used to determine the vibrations of beams with general boundary conditions. It is subsequently extended to the vibrations of two arbitrarily coupled beams, multi-span beams under moving loads, plates with arbitrary boundary supports, and built-up structures composed of any number of beams and plates. The excellent accuracy and convergence of the analytical solutions have been repeatedly demonstrated by numerical examples with varying degrees of difficulties. The improved Fourier series method actually represents a general and powerful mathematical technique for solving a wide range of boundary value problems including the vibrations of various structural components and systems.

Chapter 3 - This chapter focuses on the use of Fiber Optic-Coupled Grazing Angle Probe/Fourier Transform Reflection Absorption Infrared Spectroscopy (FOC-GAP/FT-RAIRS) as a new and promising technique for detecting and characterizing residues of neat energetic materials and their mixtures on stainless steel, glass and plastic surfaces. Smearing and TIJ techniques were used for transferring the target analytes to the substrates to be used as standards and samples. The sample transfer methods gave good sample distribution, reduced sample loss upon transfer and were easy to manipulate, giving good reproducible distributions. Samples with surface concentrations ranging from micrograms/cm^2 to nanograms/cm^2 of the explosives 2,6-dinitrotuelene (DNT), 2,4,6-trinitrotoluene (TNT), octogen (HMX), Tetryl, pentaerythritol tetranitrate (PETN), triacetone triperoxide (TATP), and PETN/2,4,6-TNT mixtures were deposited on glass, stainless steel or plastic. Methanol was used as a transfer solvent for the smearing sample preparation. Data were analyzed using chemometrics routines, specifically partial least squares (PLS). The methodology is remotely sensed *in situ* and can detect nanograms of the compounds. It is solvent free and requires neither sample preparation nor pre-treatment. The results show that detection limits as low as 10 ng/cm^2 can be detected using FOC-GAP/FT-RAIRS.

Chapter 4 - Recently, Fourier transform infrared (FTIR) microspectroscopy has been employed to detect cancer tissues from normal ones. In most of these researches the infrared spectral characteristics have been analyzed by statistical tools to study variations in metabolites. It is effective to utilize chemometric techniques for reliable modeling and classification of human tissue samples from different cancers. Principal components analysis (PCA) and analysis of variance (ANOVA) techniques are used to make an initial decision about the feasibility of microspectroscopic analysis. Multivariate data analysis procedures based on linear discriminant analysis (LDA) and soft independent modeling of class analogy (SIMCA) are also performed to classify the obtained spectra. Studying the tissue samples has drawbacks such as difficult sampling, low speed sample preparation and very sensitive keeping conditions. Also sample contamination would affect the diagnosis results while the sampling procedure from human may help the malignancy to be more developed. Thus, blood samples from healthy people and those affected by cancer are also possible to be analyzed by ATR-FTIR microspectroscopy and the obtained data would be processed by aforementioned chemometric techniques. Initial modeling is performed to classify the blood samples from healthy people and malignant cases. Then basal cell carcinoma is investigated as the case study. The obtained results confirm that FTIR microspectroscopy in combination with chemometrics would be of high interest to be introduced as a diagnostic route in medical oncology.

Chapter 5 - The study of diesel engine exhaust emissions is important due to their impact on atmospheric chemistry and air pollution. Although information on the general nature of diesel emissions is widely available, the gas and particulate phase characterization is still limited. Is known that hundreds of compounds are emitted from combustion and that they depend on type of engine, fuel composition, catalyst and engine conditions. In particular, the effects of diesel reformulation and engine operating parameters have an important impact on the gas and the particulate phase composition. Moreover, the primary diesel emissions are transformed following atmospheric degradation reactions (photolysis, photo-oxidation processes, darkness chemistry based on NO_3 and ozone). These conversions depend on initial exhaust composition, concentration of oxidants, sunlight intensity and atmospheric conditions.

In this chapter, a revision of main atmospheric transformations of diesel exhaust emissions has been performed. A comprehensive description of the monitoring strategies - dilution tunnels, on-road measurements, photoreactors, etc - has been also included. In this sense, the use of large photoreactors allows a better monitoring of different primary and secondary compounds, since precise conditions can be reproduced, isolated both from other pollution sources as other atmospheric processes. In the following section, advantages and limitations of this type of facilities have been discussed. For this, a complete description of the basic elements, the analytical instrumentation and experiment protocol has been performed. Finally, a state-of-art about diesel exhaust research in photoreactors has been also reviewed.

An example of case study - PAH experiments at European Photo-reactor (EUPHORE) - has been included. These condensed compounds are important due to their high toxicity and their correlation with gaseous compounds, chemical composition of fuels and engine operating conditions. In conclusion, the kind of information provided by photoreactors, contributes for the development of air quality programs and for the design of new alternative engines and fuels.

Chapter 6 - The Universal Attenuated Total Reflectance Fourier Transform Infrared (UATR-FTIR) equipped with a ZnSe-Diamond composite crystal allows collection of FTIR spectra directly on a sample without any special preparation. The instrument is equipped with a "pressure arm" which is used to apply a constant pressure to the cotton samples positioned on top of the ZnSe-Diamond crystal to ensure a good contact between the sample and the incident IR beam and prevent the loss of the IR beam. The amount of pressure applied is monitored by the Perkin-Elmer FTIR software. In this chapter a review of applications of the FTIR to study cotton fibers. The UATR-FTIR is used to investigate the structural changes that occur during the different developmental stages of cotton fibers from fibers initiation to maturity. The UATR-FTIR is used also to analyze contaminated cotton fibers. In this case, the Principal Component Analysis of the FTIR spectra showed that it is possible to discriminate between contaminated and non-contaminated cotton. The results obtained showed that the ZnSe-Diamond composite crystal FTIR accessory could be used as a fast and precise nondestructive method to discriminate between contaminated and non-contaminated cotton. Furthermore, the UATR-FTIR is routinely used to investigate the cellulose chemical functionalization to impart different properties to cotton fabric.

Chapter 7 - Surface-analyte parameters, which play primary roles in the preparation of samples and standards of solid energetic materials deposited on surfaces used in the development of detection technologies for these compounds, were evaluated using Grazing Angle Probe-Fiber Optic Coupled FTIR Spectroscopy and Grazing Angle Objective FTIR micro-spectroscopy. Among the properties investigated were analyte residence time, sublimation rate from surfaces, extent and homogeneity of surface coverage, phase crystallinity, surface-analyte interactions, influence of solvent and degradation of the following high explosives: 2,4-DNT, 2,4,6-TNT, Tetryl, HTMX and RDX and the homemade energetic material TATP deposited on stainless steel surfaces. Chemometrics analysis tools, such as partial least squares, were used for quantification and determination of limits of detection and quantification. The analytical measurements were related to the parameters evaluated. Homogeneities of distributions were evaluated using point mapping analysis. Statistical figures of merit were used to classify the distributions and to interpret the results obtained. Residence times for different energetic compounds deposited on surfaces were measured at several temperatures. The variability of the properties of samples prepared was found to be dependent on residence time and surface loading homogeneity. RDX and HMX were classified as very robust candidates for sample and standard preparation based on long residence times and the quality of films deposited on the surfaces.

Chapter 8 - A simple, rapid and low-cost method is presented as a new alternative for the detection of trace amounts of organic compounds on surfaces. This methodology uses optical fibers coupled to a Grazing Angle Probe-Fourier Transform Infrared Spectrometer, which allows remote sensing and direct detection of contaminants left on the surfaces of pharmaceutical reactors. This method is useful for modern programs in cleaning validation, is solvent free and requires no sample preparation. Smearing deposition was used to transfer the target analyte on the substrates to be used as samples and standards. Samples of an active pharmaceutical ingredient, provided by Bristol-Myers Squibb in Humacao Puerto Rico, and magnesium stearate, used as an excipient in concentrations ranging from 0.07 to 10.0 $\mu g/cm^2$, were deposited on stainless steel metal surfaces. Methanol was used as the transfer solvent for smearing. The amount of analyte was related to the intensity of absorption bands due to fundamental vibrations of the analyte in the fingerprint region of the mid-infrared spectra. To establish the relationship between the deposited amount of analyte and the intensity of the infrared signals, a multivariate calibration procedure using partial least squares regression and a discriminant analysis coupled with principal component analysis was used. The proposed method has a limit of detection 280 ng/cm^2 with a relative error of 3.6%.

Chapter 9 - One of the most versatile analytical chemical techniques is Fourier transform infrared (FTIR) spectroscopy. This technique is increasingly used for quantitative and qualitative analyses in many diverse applications. Among these are the analysis of pharmaceuticals, biomembranes, biopolymers and microbiological applications.

In this chapter, we have described the study of structural modification of single wall carbon nanotubes (SWCNTs) caused by γ irradiation. SWCNTs were irradiated in three different media: water, air and 30 % ammonia solution. Different techniques were used for investigation of modification of SWCNTs: Raman spectroscopy, FTIR spectroscopy and atomic force microscopy.

FTIR spectroscopy were used to examine covalent modifications of nanotubes' sidewalls. This technique has verified the functionalization of SWCNTs by detecting the presence of amino and hydroxyl groups at the sidewalls of SWCNTs.

In: Fourier Transform Infrared Spectroscopy
Editor: Oliver J. Rees, pp. 1-26

ISBN: 978-1-61668-835-6
© 2010 Nova Science Publishers, Inc.

Chapter 1

FOURIER TRANSFORM INFRARED (FTIR) SPECTROSCOPY: DEVELOPMENT, TECHNIQUES, AND APPLICATION IN THE ANALYSES OF FATS AND OILS

*Yaakob B. Che Man *, Z. A. Syahariza, and Abdul Rohman*

Halal Products Research Institute, Universiti Putra Malaysia,
43400, Selangor, Malaysia
Department of Food Technology, Faculty of Food Science and Technology, Universiti
Putra Malaysia, 43400, Selangor, Malaysia
Department of Pharmaceutical Chemistry, Fac. of Pharmacy,
Gadjah Mada University, 55281, Yogyakarta, Indonesia
Halal Products Research Institute, Universiti Putra Malaysia,
43400, Selangor, Malaysia

ABSTRACT

Infrared (IR) spectroscopy is one of the vibrational spectroscopic techniques based on the interaction between electromagnetic radiation (EMR) and matters in the IR regions. It is rapid, sensitive, non destructive, relatively simple, ease of sample presentation, and can be easily adapted to be performed by untrained users in some laboratories. In recent years, because of the development of Fourier transform infrared (FTIR) spectroscopic instrumentation, the application of this technique has increased in the study of fats and oils. The main advantages of FTIR instruments over dispersive instruments are that they have increased sensitivity (high signal to noise ratio), permitted much higher energy throughput, and considerably improved the speed of spectral acquisition. The advance of FTIR spectroscopy has attracted some scientists for qualitative analysis of fats and oils owing to the large amount of structural information that can be extracted from the infrared spectra, and because of the "fingerprints" of

---

* Tel: +603-89430405; Fax. +03-89439745, email: yaakobcm@gmail.com (Prof. Dr. Yaakob B. Che Man)

functional groups that can be displayed narrowly and intensely in the mid IR region (4000–400 cm^{-1}). Combined with improvement of computer software and chemometrics, FTIR spectroscopy can easily manipulate the spectral information and become a powerful analytical technique for quantitative analysis. In this chapter, we describe an overview of recent applications of FTIR spectroscopy in combination with chemometrics for: (1) analysis of various quality parameters of fats and oils; (2) analysis of minor components and contaminats in fats and oils; (3) authentication of fats and oils; and (4) monitoring the stability of fats and oils. The basic principles of IR absorption and some techniques related to FTIR spectroscopy are also highlighted.

INTRODUCTION

Currently, like other types of vibrational spectroscopy, IR spectroscopy represents an attractive analytical technique for analysis of fats and oils, because it is fast, non-destructive technique, sensitive, and easy in sample preparation [1] either in near infrared (NIR) or in mid-infrared regions [2]. IR spectroscopy has shown notable growth in monitoring and quality control of fats and oils caused by instrumental developments and advances in chemometrics software [3].

Since the past three decades from its development, Fourier transform infrared (FTIR) spectroscopy has revitalized the field of IR spectroscopy. This is due to not only to the better performance of FTIR spectrometers compared with dispersive IR spectrometers, but also to a number of some other factors, namely, (i) the development of FTIR spectrometers has been paralleled with advances in sample handling techniques such as the introduction of horizontal attenuated total reflectance (HATR), and (ii) the dedicated computer that is an integral part of all systems in FTIR spectrometers has been expansively exploited in the development of sophisticated data analysis [4]. As a result, FTIR spectroscopy can be used for qualitative and quantitative purposes [5].

The high sensitivity (high spectral signal to noise (S/N) ratio) obtained from FTIR spectrometers allows the detection of constituents in fats and oils, even at low concentrations such as α-tocopherol present in some oils, as well as to monitor the subtle compositional and structural differences between and among multi-constituents [3].

INFRARED (IR) SPECTROSCOPY

Spectroscopy can be defined as the interaction between electromagnetic radiation (EMR) and matters [6]. Infrared (IR) spectroscopy is a powerful technique for analyzing samples quantitatively and qualitatively, either for inorganic or organic samples in the form of gases, liquids, and solids [7]. IR is divided into three regions - the far infrared (< 25 μm) or 400-10 cm^{-1}, mid-infrared (2.5 - 25 μm) or 4000 - 400 cm^{-1} and the near infrared (<2.5 μm) or 14285-4000 cm^{-1} regions. The possibility of using FTIR spectroscopy to characterize any sample in virtually any state of matter has been realized by a judicious choice of sampling technique. The IR spectrum is generally considered as one of the more characteristic properties of a compound [8]. The IR spectrum can be obtained by passing radiation sources through a sample and determining what fraction of the incident radiation is absorbed at a particular energy. The energy at any peaks in IR spectrum corresponds to the vibrational frequency of functional groups present in the sample molecule [9].

Infrared Absorption Process

IR spectroscopy is mainly related to molecular vibrations. The molecular absorption of EMR in the IR region can cause the transitions between the rotational and vibrational energy levels of the ground (lowest) electronic energy state in the molecules [4]. As other absorption processes in spectroscopy techniques, the absorption of IR radiations is a quantized process, meaning that the functional groups in molecule samples only absorbed IR radiation at selected frequencies (energies) which corresponds to energy changes in the order of 2 to 10 kcal/mol. Radiation in this range corresponds to the stretching and bending vibrational frequencies of the bonds in the most covalent molecules.

To absorb IR radiation, bonds in the molecule must have dipole moment such as CH_2 [8]. The modes of vibration can be either stretching (change in bond length) or bending (change in bond angle). Stretching can be symmetrical (in plane) or asymmetrical (out of plane), and the bending vibration is identified as rock or deformation when moved in the same or in opposite direction, respectively.

Instrumentation

FTIR spectroscopy is based on interferometry, therefore, it differs fundamentally from traditional dispersive IR spectroscopy [10].

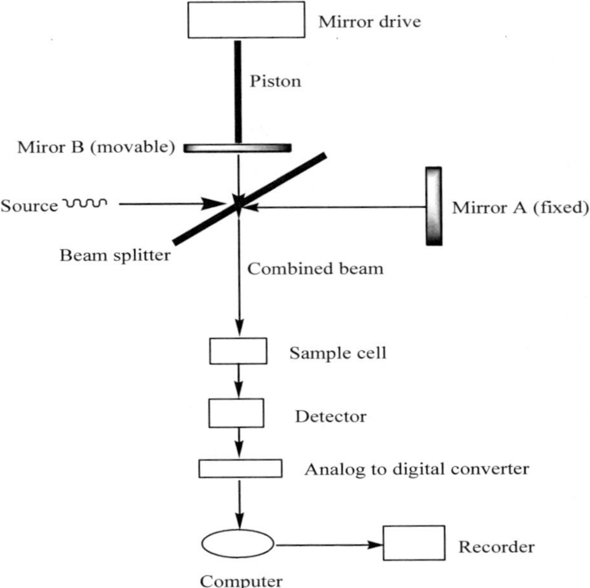

Figure 1. Schematic of an FTIR spectrometer [12].

The Michelson interferometer (Figure 1), employed in most FTIR spectrometer, uses a beam splitter, normally made up of KBr coated with germanium (for mid IR) in order to divide beam of radiation from the IR source into two parts, one part being reflected to stationary mirror and the another part being transmitted to a moving mirror [3].

When the beams are reflected back, they will recombine at the beam splitter, producing a constructive/destructive interference patters due to the varying differences between the distance traveled by two components of the beam, and part of the recombined beam subsequently reach to detector.

After the IR energy has been selectively absorbed by a sample located between the beam splitter and the detector, fluctuations in the intensity of energy reaching to detector are digitalized in real time, yielding an interferogram containing all the information required to produce the IR spectrum of the sample, but this information is in the time domain. In order to obtain a conventional IR spectrum, the interferogram is converted to the frequency domain using Fourier transformation, a mathematical algorithm used to decodify the interferogram and to obtain interpretable information about the individual frequencies [11].

IR Spectra of Fats and Oils

The principal components composing of fats and oils are triglycerides, with different substitution patterns, lengths, and degrees of saturation of the chains. Besides, fats and oils also contain some minor components such as phospholipids and tocopherols.

Although spectra of fats and oils look to be almost similar, they differ in the intensity of their bands as well as in the exact frequency at which the maximum absorbance is produced in each case, caused by the different nature and composition of the studied fats and oils [13]. These differences are exploited to discriminate among fats and oils. Figure 2 is a spectrum example of extra virgin olive oil. The peak assignments and other peaks which are possible to appear in edible fats and oils are described by Guillen and Cabo [15].

Table 1 describes the frequencies of the band and shoulder which are characteristics of fats and oils. Due to the overlapping in the bending vibrations modes, the assignment of bands corresponding to the stretching vibration modes is easier than that of bands corresponding to the bending vibration modes.

Table 1. The frequencies of band and shoulder of edible fats and oils [15]

Frequency (cm^{-1})	Functional group	Mode of vibration	Intensity
3468 (band)	-C=O (estser)	Overtone	weak
3450 (band	-OH (hidroxy)	Stretching	Very strong
3025 (shoulder)	=C-H (*trans*)	Stretching	Very weak
3006 (band)	=C-H (*cis*)	Stretching	medium
2953 (shoulder)	-CH (methyl)	Stretching (asymmetric)	medium
2924 (band)	-CH (methylene)	Stretching (asymmetric)	Very strong
2853 (band)	-CH (methylene)	Stretching (symmetric)	Very strong
2730 (band)	-C=O (estser)	Fermi resonance	Very weak
1746 (band)	-C=O (estser)	Stretching	Very strong

1711 (shoulder)	-C=O (acid)	Stretching	Very weak
1654 (band)	-C=C (*cis*)	Stretching	Very weak
1648 (band)	-C=C (*cis*)	Stretching	Very weak
1465 (band)	-CH (methylene; methyl)	Bending (scissoring)	medium
1418 (band)	=C-H (*cis* disubstituted olefins)	Bending (rocking)	weak
1400 (band)	Difficult to be assigned	Bending	weak
1377 (band)	-CH (methyl)	Bending (symmetric)	medium
1319 (band; shoulder)	Difficult to be assigned	Bending	Very weak
1238 (band)	-C-O; -CH$_2$-	Stretching; bending	medium
1163 (band)	-C-O; -CH$_2$-	Stretching; bending	strong
1118 (band)	-C-O	Stretching	medium
1097 (band)	-C-O	Stretching	medium
1033 (shoulder)	-C-O	Stretching	Very weak
968 (band)	-HC=CH-(*trans*)	Out of plane	weak
914 (band)	-HC=CH-(*cis*)	Out of plane	Very weak
723 (band)	-(CH$_2$)$_n$; -HC=CH- (*cis*)	Bending (rocking)	medium

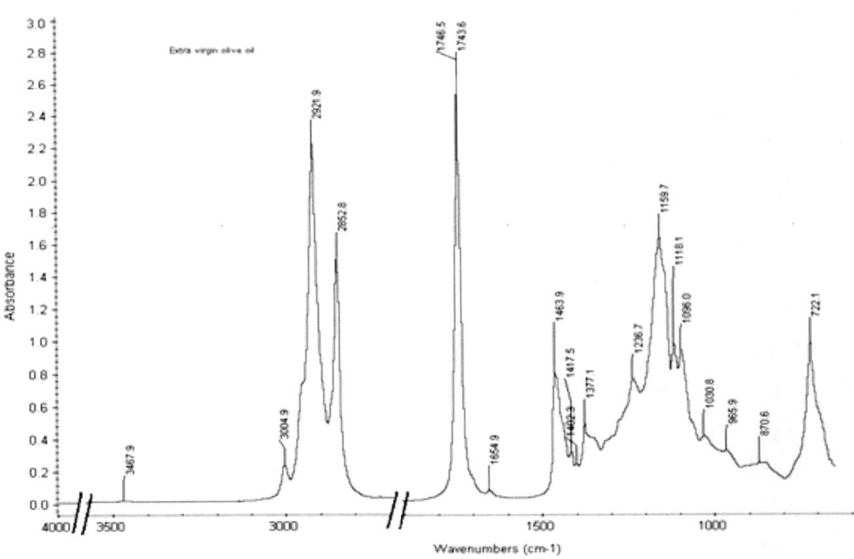

Figure 2. FTIR spectrum of EVOO obtained at resolution 4 cm^{-1} and co-scan 32 [14].

Sample Handling Technique

One of the critical aspects in IR spectroscopy is related to sample preparation. Because of the large diversity uses of IR spectroscopy in analyzing and characterizing of samples, a large number of sample preparation techniques have been developed over the years, as some spectroscopists have striven to produce better quality spectra effectively and efficiently. A large number of accessories are available for determining the samples, either qualitatively or quantitatively, in various states (gases, liquids, or solids). They can be divided into a few categories; namely transmission methods, internal reflectance, external reflectance, diffuse reflectance, photoacoustic detection, and GC/IR [16]. Several publications contain the detailed descriptions of the various sample handling techniques used in mid IR can be met elsewhere. Among those techniques, the two most important sample presentation methods are transmission technique and attenuated total reflectance (ATR). Both methods require minimum sample preparation.

1. Transmission Technique

Despite the apparent advantages of attenuated total refectance (ATR) methods, transmission-based applications are still widely used. However, sample pre-treatment of this kind is nearly always exploited [17]. Transmission cell is a commonly used sample handling device, especially in mid IR and can be applied to samples in gases, solids, and liquids. The IR beam passes through sample, and only the radiation which is not absorbed by the sample reach the detector [18].

Liquid transmission cells can be met in several forms; sealed, demountable, variable pathlength, microcavity, and variable temperatures. The demountable cell can be used in a fixed pathlength mode as well. It can be disassembled for cleaning, and the spacer can be changed to provide flexibility in pathlength choice. The sealed liquid cells are fixed pathlength. It is recommended to periodically measure the cell thickness, since some window materials slowly dissolve with usage [19]. The spacer which is located between the two windows will determine the sample thickness or cell pathlength.

Most IR windows are highly polished salt of various types of crystals. The most common encountered and least expensive window materials are sodium chloride (NaCl) and potassium bromide (KBr). Both have good transmission characteristics across most of the mid-IR spectral range. Unfortunately, both these materials are highly water soluble, therefore, NaCl and KBr will be destroyed by aqueous samples. To overcome this problem, the window materials composed of CaF_2 and BaF_2 cells are commonly employed with aqueous samples [4].

2. Attenuated Total Reflectance (ATR)

ATR is one of the Internal Reflection Spectroscopy (IRE), in which the sample is positioned in contact against a special crystal called an internal reflectance element (IRE). The IRE should be made from a material with high reflection index, such as zinc selenide (ZnSe) and germanium (Ge) [20].

The ATR technique provides a simple and convenient means of acquiring the IR spectra of a wide variety of sample types, many of which are not readily convenient to IR analysis using conventional transmission measurements [21].

In order to attain the success of ATR, the following two requirements must be reached: (i) the sample must be in direct contact with the ATR crystal, because the evanescent wave or bubble only extends beyond the crystal 0.5 μ - 5 μ; (ii) The refractive index of the ATR crystal must be significantly greater than that of the sample, otherwise internal reflectance will not occur, consequently the light will be transmitted rather than internally reflected in the crystal. Usually, refractive index values of ATR crystals are between 2.38 and 4.01 at 2000 cm^{-1} [22].

To minimize the random errors and to verify the measurement reproducibility, a minimum of two complete sampling should be made (duplication), because multiple examinations verify the homogeneity and minimize errors [23]. Different sampling techniques have been used in obtaining better quality spectra, and new sensitive techniques have been developed in order to examine formerly intractable samples [7; 11; 17].

Chemometric Data Analysis

One of the important factors contributing to the success of FTIR analysis as a quantitative tool is chemometric package. Chemometric is an approach to analytical and measurement science based on the idea of indirect observation [24].

Chemometrics is a discipline of extracting chemically relevant information from data produced in chemical experiments by means of statistical and mathematical tools [1; 25]. It is an indirect approach to the study of the effects of multivariate factors (or variables) and hidden patterns in a complex set of data [26].

Chemometric data analysis has the advantage that it is able to deal efficiently with real-world multivariate data, including highly co-linear of spectral data.

Many statistical procedures in multivariate statistics such as multivariate calibrations, classification, and cluster analysis have been used in solving the authentication problems of fats and oils [27]. A number of reviews and papers on the use of chemometrics in food analysis had explained the importance of this method in assessing problems in food analysis especially in authenticity of foods, including fats and oils, as reported by Arvanitoyannis and his colleagues [28; 29; 30] and de la Fuente and Juarez [31].

Multivariate calibration is one of useful chemometric techniques and often used for analysis of fats and oils, as it facilitates the rapid and simultaneous determination of each component in the complex mixtures without time consuming and with minimal sample preparation [32]. The most commonly used multivariate calibration methods are partial least square (PLS) and principal component regression (PCR). Both calibration methods are based on reduction of spectral data and inverse calibration. Using these, it is possible to make a calibration for the desired component while completely modeling the other source of variations [33].

PLS is able to use information of sample spectra from large spectral frequencies and subesequently correlate the changes of spectral absorption with its analyte concentration, whilst concomitantly computing te other spectra which may perturb the analyte spectra [34]. Meanwhile, PCR is a type of factor analyses, in which the spectral and concentration data are incorporated into the model in one step [35].

Another chemometrics technique commonly used especially for authentication of specific oils from others is discriminant analysis, one of the classification methods which finds the relationships between a set of descriptive variables (FTIR spectra) and a qualitative variable (class of fats and oils) [365]. This chemometrics was successfully employed for classification of VCO from other vegetable oils [37] and cod liver oil from animal fats [38].

APPLICATION OF FTIR SPECTROSCOPY IN FATS AND OILS ANALYSES

Since from its introduction, FTIR spectroscopy has evolved to become an important tool for quantitative analysis. Its application in the qualitative and quantitative analyses of edible oils has been an active area of research [39]. FTIR spectroscopy has played a major role in fundamental research on lipid systems [18]. In terms of practical application, it has been broadly used in the industry of fats and oils to monitor the quality of fats and oils for the purpose of quality control.

Analysis of Various Quality Parameters of Fats and Oils

Typical analyses that are important for monitoring various quality parameters of fats and oils are usually based on standard test method from the American Oil Chemists' Society (AOCS) and International Union of Pure and Applied Chemistry (IUPAC) [40]. These test methods require the use of toxic chemical, laborious procedures, and time consuming. For this reason, FTIR is proposed to be alterative methods for AOCS and IUPAC methods. Previous reports have successfully revealed the power of FTIR spectroscopy as rapid and safe alternative tools in determining and monitoring the quality parameters of fats and oils. Some are listed in Table 2.

1. Iodine value and Saponification Value
Iodine value (IV) is an important parameter in oil and fat specifications. IV has been used to predict the chemical and physical properties of fats and oils, such as oxidative stability and melting point [41].

Table 2. List of published works related to the use of FTIR spectroscopy for the monitoring the quality of fats and oils

Parameter	Oils/fats matrices	Sample handling device	Chemometrics	References
Iodine value	Eleven vegetable oils	NaCl transmission cell	-	[46]
	almond, castor, corn, coconut, cottonseed, linseed, olive, palm, peanut, safflower, seed, sesame, soybean, sunflower, and wheat germ	NaCl transmission cell	-	[47]
	RBD soybean oil	single-bounce ATR[a]	PLS[b]	[48]
	Palm oil and its products	BaF2 transmission flow cell.	PLS	[49]
	Palm oil	BaF2 transmission flow cell.	PLS	[50]
	hydrogenated rapeseed and soybean oils	KCl flow cell	PLS	[51]
	Standard triglycerides	ATR	PLS	[52]
Saponification number	Standard triglycerides	ATR	PLS	[52]
FFA	Virgin olive oil	ATR	PLS	[32]
	Palm olein	NaCl transmission cell	PLS	[53]
	RBD palm olein	BaF2 heated flow cell	PLS	[54]

	Olive oil	HATR	PLS	[55]
	Feed lipid	single bound-ATR	PLS	[56]
Peroxide value	sunflower oil, safflower oil, and rapeseed oil	KBr transmission cell	-	[57]
	Sufflower oil	CsI transmission cell	-	[58]
	palm olein	NaCl transmission cell	PLS	[59]
	Crude palm oil and crude palm kernel oil	NaCl transmission cell	PLS	[60]
	Edible oils	KCl heated flow cell and and 3M IR card	PLS	[61]
	olive, corn, soy, sunflower, canola and peanut	CaF$_2$ flow cell	PLS	[62]
Anisidine value	sunflower oil, safflower oil, and rapeseed oil	KBr transmission cell	-	[57]
	Thermally Oxidized Palm Olein	NaCl transmission cell	PLS	[63]
Hydroxyl value	Soybean oil	horizontal attenuated total reflectance (HATR)	PLS	[64]
Thiobarbituric acid reactive substances	Palm olein	NaCl transmission cell	PLS	[65]
	Palm olein	NaCl transmission cell	PLS	[66]
	Palm olein	NaCl transmission cell	PLS, PCR^c	[67]
Peroxide value	sunflower oil, safflower oil, and rapeseed oil	KBr transmission cell	-	[57]
trans content	Bakery products	ATR	-	[68]
	RBD soybean oil	single-bounce ATR	PLS	[48]
	hydrogenated rapeseed and soybean oils	KCl flow cell	PLS	[51]
	Partially hydrogenated fat	ATR	-	[69]
	Canola oil	NaCl transmission cell	PLS	[70]
	Soybean oil	NaCl transmission cell		[71]
	hydrogenated soybean oil	NaCl transmission cell	-	[72]
	Shortening and edible oils	NaCl transmission cell	-	[72]
Moisture content	Palm olein	ATR	PLS, PCR	[74]
	Crude palm oil	NaCl transmission cell	PLS	[75]
Slip melting point	Palm oil	ATR	PLS, PCR	[76]
cloud point	Palm oil	ATR	PLS, PCR	[77]
Solid fat index	butter, soybean oil, and lard	ATR and PAS	-	[78]

^{a}ATR, attenuated total reflectance; ^{b}PLS, partial least square; ^{c}PCR, principle component regression; -, not reported.

IV is often used as a screening guide to check adulteration of crude palm and crude palm kernel oil. It measures degree of unsaturation in fats and oils, and is expressed as the number of grams of iodine absorbed by 100 gram of oil under the test conditions. Currently, IV of oil is evaluated by the Wijs method according to AOCS Cd I-25 [42] or ISO 3961 [43].

Saponification number (SN) is related to the average molecular weight of the triglycerides in the mixture and is determined by the amount of KOH necessary to saponify a certain amount of fats or oils. High SN indicates the small average molecular weight and vice versa [44]. As part of the efforts to eliminate highly toxic solvents and to be environmental friendly, FTIR study has been carried out to determine IV and SN.

Van de Voort et al. [45] have reported the use of an ATR- FTIR to determine IV and SN for oils and fats. Calibration standards were prepared using commercially available triglycerides. PLS treatment was used to determine the spectral correlation between the known chemical IV and SN values with the calibration model. Result revealed that region at 3200 – 2600, 1600 – 1000, and 1850-1000 cm^{-1} were optimal for the determination of IV and SN, respectively.

Che Man et al. [50] have successfully used FTIR spectroscopy to determine IV of palm oil products. A calibration standard was developed by blending palm stearin and super olein in specific ratios that covered a range of 27.9 to 65.3 IV units. The region between 3050 and 2984 cm^{-1} was used to construct a calibration model using PLS treatment. A validation approach was used to optimize the calibration model with R^2 of 0.9995 and standard error of prediction (SEP) of 0.151, respectively. In another study, combined with multivariate calibration, Che Man and Setiowaty [49] had developed FTIR technique for determination of IV in palm oil and its products. Forty-two spectra of palm oil with IV ranging between 53 and 65 were used to develop calibration model. The spectra of these standards were measured in the range between 3025 and 2992 cm^{-1}, corresponding to the absorption band of *cis* C=CH stretching vibration. Multivariate calibration technique, namely PLS and PCR, using different baseline types were applied to compare SEP, standard error of calibration (SEC), coefficient of determination (R^2) and F-ratios between both techniques.

Setiowaty and Che Man [74] have developed FTIR technique for determination of SN for palm oil products. The SN of melted palm olein and palm kernel olein mixtures was pre-analyzed using AOCS chemical method. FTIR spectra were acquired using a 100 μm NaCl transmission flow cell. PLS and PCR calibrations were developed from selected spectral regions which are associated with the CH$_3$, CH$_2$, and C=O regions of the spectrum and subsequently correlated to the chemical SN values obtained using AOCS method. The accuracy of FTIR method for the determination of SN in the evaluated palm oil samples expressed as prediction error for PLS and PCR methods was 1.17% and 1.97%, respectively.

The infrared derivative spectrum of pure vegetable oils was used for determination of IV in twelve edible vegetable oils as well as castor and linseed oils. Oils with IV ranging from 10 to 190 were evaluated and found to give satisfactory values. Results with good precision and accuracy were achieved using this method, typically with relative standard deviation (RSD) of 5% and below [47].

2. Free Fatty Acid

From quality point of view, the content of free fatty acids (FFA) is one of the most important factors determining the quality and economic value of edible oils. Several methods have been described for the determination of FFA in oils. The most commonly methods used is that recommended by the AOCS [42], based on titration of the acidity against an alkali using phenolphthalein as an indicator. FFA is more likely prone to oxidation than triglycerides, so that its presence in oils increases the possibility of producing rancidity. Since the economic value of oils depends heavily on the FFA concentration, it would be desirable to measure accurately its concentration during manufacturing and storage. Hence, the importance of fast and accurate methods for fats and oils analyses has been recently stressed [79].

ATR-FTIR Spectroscopy has been applied to determine FFA content in commercial olive oil samples from different origin [79]. Different methods of selecting the training set were applied and compared. The prediction capabilities of PLS, net analyte signal (NAS) preprocessing followed by PLS or classical least squares (CLS) regression method of ATR-FTIR data were evaluated. Both direct and indirect FTIR spectroscopic methods were developed by Ismail et al. [80] for the determination of FFA in fats and oils based on transmission and ATR approaches, covering an analytical range of 0.2-8 % FFA. Both methods showed comparable results with AOCS titration method.

FTIR spectroscopy employing a portable variable filter array (VFA)-IR spectrometer equipped with a transmission flow cell was used to quantitatively analyze FFA in edible oils. FFA determination employed was based on the extraction of FFAs into methanol containing the base NaHNCN, which converts the FFA to its salts, followed by absorbance measurement of carboxylate group at frequency 1,573 cm^{-1} [81]. FTIR using single bound-ATR was also employed for analysis FFA in feed lipid [56], in edible oils using spectral reconstitution [82], in fish oils [83], and in virgin olive oil using PLS calibration [32].

FTIR spectroscopic methods also have been used to determine FFA in crude palm oil and RBD palm olein using a 100µ BaF_2 cell [54]. The samples were prepared by hydrolyzing oil with enzyme in an incubator and subsequently applied to heated sample handling accessory with a specific design for palm oil analysis. PLS method was used to construct a calibration model using FTIR carboxyl region (C=O) from 1722 to 1690 cm^{-1}. From the results obtained, it can be concluded that FTIR spectroscopy in conjunction with a flow-through transmission cell is a useful technique in determining FFA in palm oil until less than 0.1%.

Che Man and Setiowaty [53] developed a PLS calibration model for the prediction of FFA contents in palm olein, based on the spectral region of 1728 – 1662 cm^{-1}. R^2 and SE were 0.997 and 0.017 of FFA unit, respectively. This model was tested by cross-validation steps to minimize SE of the model. The method was tested for accuracy and repeatability compared to the AOCS titration method. The new FTIR spectroscopic method provides a practical procedure that can be implemented in a straightforward manner for quality control applications in the fats and oils industry.

In another study on FFA which is based on the different profiles of fatty acid, Verleyen et al. [84] described the use of FTIR to determine FFA value in six vegetable oils (corn, soybean, sunflower, palm, palm kernel and coconut oils). The results showed that for those oils, up to FFA level of 6.5% for coconut oil, the best correlation coefficient was obtained by linear regression of the free carboxyl absorption at 1711 cm^{-1}.·

3. Peroxide Value (PV)

The AOCS official iodometric method for PV determination is applicable to all normal fats and oils. The standard AOCS method is based on the release of molecular iodine (I_2) by the presence of hydroperoxides stoichiometrically, when exposed to potassium iodide (KI) in an acidic environment [85]; the reaction converts the hydroperoxide to alcohols. The molecular iodine released is complexed with soluble starch, which acts as an indicator, and the iodine is quantified by titration with sodium thiosulphate ($Na_2S_2O_3$). Based on the stoichiometry of the two reactions, the hydroperoxide concentration can be calculated and is commonly expressed as miliequivalents of hydroperoxide per kilogram of fat (meq/kg).

A primary FTIR spectroscopic method for determination of PV in edible oils was developed by Ma et al. [61] based on the stoichiometric reaction of triphenylphosphine (TPP) with hydroperoxides to produce triphenylphosphine oxide (TPPO). Accurate quantization of the TPPO formed in this reaction is performed by measurement of its intense absorption band at 542 cm^{-1}. Concentration of TPPO, expressed in term of PV, ranging from 0-15 PV was used as the calibration standard. Van de Voort et al. [62] had carried out a study on the use of FTIR spectroscopy in determining PV. *Tert*-butyl hydroperoxide (TBHP) was added to peroxide-free canola oil to develop a calibration standard. The method was conducted using 1 mm CaF$_2$ flow cell and treated with PLS method. Region 3750-3150cm^{-1} was selected for calibration, which was appearing due to -OO-H stretching vibration. This method was found to be well suited in predicting PV over conventional chemical method.

4. Anisidine Value and Aldehydes

Several methods developed for measuring the content of carbonyl compounds have been quite sensitive, quantitative, and well correlated to compounds associated with the development of rancidity. The AOCS anisidine value (AnV) is a combined measure of mostly 2-alkenals and 2, 4-dienals and, to a more limited degree, saturated aldehydes, because ultraviolet (UV) absorption of the *p*-anisidine/ aldehyde reaction products varies with the aldehyde type [86]. Dubois et al. [70] had reported the simultaneous monitoring of aldehyde formation and the determination of AnV in thermally stressed oils. The study was conducted to develop a 'universal' synthetic calibration for AnV determination of oils, including the ability to analyze the types of individual aldehydes that contribute to the overall AnV. The excellent results obtained indicate that the calibration approach works well.

A FTIR spectroscopy coupled with transmission cell was described to predict AnV of a thermally oxidized palm olein [63]. In order to construct a calibration model, PLS regression technique was employed based on spectral region at 2747 – 2619 cm^{-1} and 1715 – 1673 cm^{-1} either individually or simultaneously. The precision of this method was shown to be comparable to that of the AOCS method used for measurement of AnV, with R^2 of 0.99.

5. Thiobarbituric Acid Reactive Substances

FTIR spectroscopy had been used as a non-chemical method for the determination of secondary oxidation products in edible oils [45; 13]. Malonaldehyde occurs as an end product of oxidized polyunsaturated fatty acids in fats and oils. The malonaldehyde will react with the thiobarbituric acid (TBA) reagent to produce a red colored complex with a distinctive red color having maximum absorption at λ 532 nm [87].

Malonaldehyde as one of the thiobarbituric acid reactive substances (TBARS) in palm oil and its products has been successfully determined by FTIR spectroscopy coupled with multivariate calibration [74]. In the study, the samples were oxidized by heating up to 180°C with 3 hr intervals to generate TBA in the range of 0.57 to 0.14. These samples were used in the calibration and validation steps. The calibration models based on PLS and PCR analysis were constructed using the spectral region between 1710 to 1670 cm^{-1} and 2790 to 2670 cm^{-1} and using chemical data of the calibration set. Evaluation of the calibration models was carried out by validation step. The SEP gained from the validation equation for PLS and PCR methods were in order 982 to 5516 (10^{-6}), respectively.

6. Isolated Trans Content

The presence of *trans* fatty acids (TFA) has attracted health professionals because of their association with heart disease [88]. IR spectroscopy is a widely used procedure to determine total TFA and it has been standardized by IUPAC and AOCS methods [89; 73]. The problem encountered in accurate determination of low level of *trans* unsaturation has been resolved by Ulbert and Haider [70]. The authors had applied a spectral subtraction technique, based on the '*background corrected absorptivity*'. The spectrum of fat samples was subtracted with TFA free oil samples. The determined TFA content was comparable with AOCS peak area obtained from gas chromatography method and had improved the accuracy of FTIR technique.

Sedman et al. [90] modified the standard AOCS method to make a direct rapid determination of isolated TFA content of neat fats and oils using FTIR spectroscopy. The single-beam spectra of calibration standards were subtracted against that of the base oil, eliminating the spectral interference caused by underlying triglyceride absorption, and facilitating direct peak height measurements as described in AOCS for determination of TFA content using FTIR spectroscopy method.

A rapid determination of total *trans* content in neat hydrogenated oil by ATR technique has been described by Mossoba et al. [69]. The band of C-H OOP (out of plane) deformation, observed at 966 cm^{-1} is the unique characteristic of isolated double bonds with *trans* configuration. The use of ATR cell facilitates the rapidity and convenience of FTIR method. A rapid technique for determining total fat and TFA contents in bakery products using microwave-assisted Soxhlet extraction, in combination with ATR cell was demonstrated by Capote et al. [68]. This proposed method gives similar result with Folch procedure, with a few advantages in term of time saving, solvent reduction, efficiency and cheaper alternatives.

7. Moisture Content

The rapid analysis of moisture content is crucial in the fats and oils processing for the adjustment of manufacturing processes while production is underway. In palm oil industry, palm olein is regulated in terms of its moisture content and requires continual monitoring to ensure that specifications are being met. A simple, rapid and direct FTIR spectroscopic method was developed for the determination of moisture content in crude palm oil at region of 3074-3700 cm^{-1} [75]. A PLS regression technique was employed for developing calibration model followed by cross-validation step. The accuracy of this method was comparable to that of the AOCS vacuum oven method used for determination of moisture and volatile matters with R^2 and SEC 0.978 and 0.91, respectively.

In another study, FTIR-ATR spectroscopy was used to carry out multi-component analysis in association with multivariate calibration of PLS and PCR for the determination of moisture content [74]. The FTIR spectra of mixture of palm olein and water in propanol at region of 3025 and 2992 cm^{-1}, were used to compare different multivariate calibration techniques for quantitative moisture content determination. Thirty-seven spectra of palm oil with slip melting point ranging between 0.1 and 0.5% were used to create calibration models based on PLS and PCR methods. The calibration models generated the number of factors from 3 to 7, R^2 of 0.9943 to 0.9947, SEC of 0.000672 to 0.000697 and SEP ratios of 0.000705 to 0.000706.

8. Slip Melting Point and Cloud Point

In determining slip melting point (SMP) of palm oil, a calibration standard was prepared by randomly blending palm stearin and palm olein covering a range of 22 to 55 °C [76]. The spectral region of these standards between 2992-2832 cm^{-1} corresponding to the absorption band of the CH_2/CH_3 stretching region and 1750-1724 cm^{-1}, related to the ester linkage carbonyl absorption band was used to build the PLS and PCR calibration models for the prediction of SMP. The cross-validation approaches were used to optimize the calibration and yield satisfactory R^2 (> 0.99) and the minimal SEP.

The application of FTIR spectroscopy in combination of transmission cell was also described by Setiowaty and Che Man [77] to predict cloud point of palm oil samples. The calibration set was prepared by mixing the palm olein and palm stearin in random amount (w/w) covering a wide range of cloud point. The PLS and PCR techniques were employed to construct the calibration model. These models were further accomplished by cross-validation step. The SEPs found were 0.1241 and 0.1514 for PCR and PLS, respectively. The precision of this method was shown to be comparable with conventional AOCS method used for measurement of cloud point with R^2 of 0.9923 and 0.9922 for PCR and PLS, respectively.

9. Solid Fat Index

Yang and Irudayaraj [78] has carried out the characterization of semisolid fats and edible oils by FTIR-photoacoustic spectroscopy (FTIR-PAS). The study highlights the potential use of FTIR-PAS for fats and oils analysis. The result obtained was validated with ATR technique. The FTIR-PAS has advantages in sample preparation and depth profiling. A good quality of PAS spectra can be obtained using some parameters, i.e at scanning rate of 5 kHz, 256 scan/sample and resolution of 4 cm^{-1}. The results had proved that FTIR– PAS method has a great potential for analyzing oils, solids and low moisture samples.

Analysis of Minor Components and Contaminants in Fats and Oils

FTIR spectroscopy has been employed for analysis of minor components present in fats and oils, such as β–carotene in crude palm oil [91], sesamol in sesame oil [92], and gossypol in cottonseed oil [93] and contaminants in fats and oils such as aflatoxin [94]. β–carotene is a predominant carotenoids present in crude palm oil and it is known for its provitamin A (retinol) activities. Separate PLS calibration model was developed for predicting β–carotene at regions of 976 to 926 cm^{-1} and 546 to 819 nm for FTIR and NIR spectroscopy, respectively. Both techniques provide alternative means to measure β–carotene in crude palm oil.

Sesamol may present in trace amount in sesame oil as phenolic antioxidant factors, and usually deliberated during refining of unroasted sesame seed. FTIR spectroscopy based on NaCl transmission cell accessory at room temperature combined with PLS regression statistical method was used to derive calibration models for determining sesamol spiked in sesame, refined-bleached-deodorized (RBD) palm olein and groundnut oils, with R^2 of

0.9947, 0.9940 and 0.9662 for the mentioned oils, respectively [92]. The calibration models were validated and the R^2 of validation and the SEP computed. The SDD for repeatability and accuracy of the FTIR method was comparable to the actual values of sesamol content spiked in sesame, RBD palm olein and groundnut oils. These observations strengthen the premise that FTIR spectroscopy is an efficient and accurate method for determining minor components in edible oils such as sesamol.

Analysis of gossypol using FTIR spectroscopy was performed using NaCl transmission cell. The wavelengths were selected for the gossypol and standards were prepared by spiking some clean cottonseed oil samples with the gossypol at concentrations of 0 – 5%. A PLS regression was used to derive the calibration models. The R^2 values of calibration models were computed for the FTIR predicted values versus chemical values of gossypol in percent. Based on the results obtained, the FTIR spectroscopy can be a useful instrumental method for determining gossypol in cottonseed oil [93].

Mirghani et al. [94] developed FTIR spectroscopy with ATR as a new method for the determination of aflatoxins in groundnut and groundnut cake. The wavelengths were selected for the four types of aflatoxins (B_1, B_2, G_1, and G_2) and the standards prepared for each by spiking some clean sample with the aflatoxins in concentrations of 0 -1200 parts per billion. A PLS regression was used to derive the calibration models for each toxin. The R^2 of the calibration model were computed for the FTIR spectroscopy predicted values vs. actual values of aflatoxins in parts per billion. The R^2 was found to be 0.9911, 0.9859, 0.9986, and 0.9789 for aflatoxins B_1, B_2, G_1, and G_2, respectively.

In order to retard the oxidations of oils, some natural and synthetic antioxidants were used. FTIR spectroscopy combined with chemometrics using NaCl transmission cell has been successfully used for determination of antioxidants in some oils, namely α-tocopherol in refined-bleached-deodorized palm olein [95], *tert*-butylhydroquinone (TBHQ), butylated hydroxyanisole (BHA), and prophyl gallate [95; 96; 97]. Tocopherols are one of the important minor components present in fats and oils because of its function as an effective antioxidant in lipid and lipid-containing foods. Recently, FTIR technique was developed by Che Man et al. [95] for the determination of α -tocopherol in RBD palm olein. The calibration and validation samples were prepared by spiking known amount of α -tocopherol up to 2000 ppm in the sample. Spectral region at 3100-2750cm^{-1} was used to construct a calibration model using PLS treatment. It was found that this method was comparable to HPLC analysis, with R^2 of 0.9922.

Authentication of Fats and Oils

Because of its ability to provide fingerprint, IR spectroscopy has been identified as an ideal analytical method for authenticity studies of edible fats and oils [99] because it gives valuable information about the presence of molecular bonds in the evaluated samples [8]. Since very early time, oils and fats have been liable to adulteration, either intentionally or accidentally to a greater or lesser degree. Adulteration is increasingly more difficult to be detected when the adulterant has a chemical composition which is similar to that of the original oil [100]. Combined with suitable chemometrics, FTIR spectroscopy can provide a powerful technique for authentication of fats and oils. Some papers related to authentication of edible fats and oils are listed in Table 3.

As a relative new comer in oil industry, virgin coconut oil (VCO) can be subjected to adulteration with cheaper oils like palm kernel oil (PKO) and palm oil (PO). FTIR spectroscopy combined with PLS and discriminant analysis was succesfully used for quantification and classification analyses of VCO with PKO and PO, respectively [37, 101].

Due to the high price value compared with other oils, extra virgin olive oil (EVOO) can be a target of adulteration to gain economical profits. For this reason, several researches have developed FTIR spectroscopy for analysis of some adulterans in EVOO.

Lai et al. [26] had investigated the potential of FTIR spectroscopy, in conjunction with PCA and discriminant analysis for determining extra virgin and refined olive oil authenticity.

The application of PCA on the spectra had successfully clustered the samples according to plant species. 93% of samples in calibration set and 100% samples in a validation set were correctly classified after subjected to discriminant analysis. Therefore, FTIR technique has been shown to successfully discriminate among different vegetable oils, and types of olive oil in particular after establishment of complete database comprising spectra from all possible species.

FTIR spectroscopy is also used as a rapid technique for the quantitative determination of adulterants in extra virgin olive oil, namely refined olive and walnut oil [115]. The method was carried out in combination with ATR and PLS treatment.

A series of mixtures of extra virgin and refined olive oil with known concentration is analytically prepared and used as the calibration standard. After PLS treatment, the best model was used to determine the concentration of samples in the independent validation set. The results obtained clearly demonstrated the potential of this technique to determine the addition of low levels of refined olive oils.

Table 3. List of published works related the use of FTIR spectroscopy in the authentication study of fats and oils

Oils	Adulterant	Chemometrics	References
Extra virgin olive oil	sunflower, corn, soybean and hazelnut	MLR[a] and LDA[b]	[102]
	sunflower, corn	PCA[c], PLS-DA[d]	[103]
	corn, sunflower and soybean	PLS	[104]
	sunflower oil, soyabean oil, sesame oil, corn oil	PLS	[105]
	Sunflower	HCA[e], SIMCA[f], PLS	[106]
	Olive pomace oil	PLS	[107]
	refined olive, sunflower, rapeseed, peanut, soybean and corn oil	PLS, LDA[g], ANN[h]	[108]
Olive Oil	sunflower oil, rapeseed oil, and soybean oil	PLS and PCA	[109]
	soya oil, sun flower oil, corn oil, walnut oil and hazelnut oil.	PLS	[110]

	Canola, walnut, sunflower, soybean, peanut, corn, Sesame	PLS, DA[i]	[111]
Virgin coconut oil	Palm oil	PLS, DA	[101]
	Palm kernel oil	PLS, DA	[37]
camellia oil	soybean oil	SIMCA, PCA, PLS	[112]
Refined olive oil	Hazelmut oil	SLDA[j]	[113]
Shortening in cake	Lard	PLS	[114]
Cocoa butter in cake formulation	Lard	PLS	[34]
Cod-liver oil	Animal fats (lard, body fats of lamb, cow, and chicken)	PLS, DA	[38]

[a]MLR, multiple linier regression; [b]LDA, linier discriminant analysis; [c]PCA, principal component analysis; [d]PLS-DA, partial least square-discriminant analysis; PLS, partial least square; [e]HCA, hierarchical cluster analysis; [f]SIMCA, soft independent modeling of class analogy; [g]LDA, linear discriminant analysis; [h]ANN, artificial neural network; [i]DA, discriminant analysis; [j]SLDA, stepwise linear discriminant analysis.

Monitoring Oxidation of Vegetable Oils

Due to oxidative deterioration, fats and oils can become rancid, which causes a major problem in food industry. Edible fats and oils suffer from thermal degradation, when they are subjected to high temperatures in a constant manner, and because of that, it is interesting to evaluate the heating conditions applied to edible oils. Numerous methods to determine the rate of oxidation processes are related to the measurement of the concentration of primary or secondary oxidation products or both, or to the amount of oxygen consumed during the oxidation process [116].

Because of its ability to provide molecular structure information, FTIR spectroscopy has also been considered to be a useful technique for the rapid measurement of lipid oxidation [117], either using sequentially recording the spectra of samples taken over time from edible fats and oils under oxidative stress or using continuously monitoring spectral changes of edible fats and oils heated on ATR surfaces [85;13]. Van de Voort *et al.* [118] had conducted a study on monitoring the oxidation of edible oils, namely safflower and cottonseed oils using FTIR spectroscopy. These oils were oxidized under various conditions, and their spectral changes were recorded, identified, and interpreted. The absorption spectra associated with common oxidation were identified and related to the reference compounds. The power and usefulness of FTIR spectroscopy to monitor oxidative changes were demonstrated through the use of "real-time oxidation plots".

Guillen and Cabo [57] have used FTIR spectroscopy for monitoring oxidative stability of different edible oils, including safflower, sunflower, rapeseed and olive oils. These oils were subjected to oxidation in a convection oven with air circulating at around 70 °C. FTIR spectroscopy was employed to monitor oxidation processes by collecting spectra from time to time. The changes observed in IR spectra can be used as useful indicators of oxidative stability of edible oils and are closely related to changes observed in peroxide and anisidine values in the course of the oxidation of the samples. The authors concluded that FTIR spectroscopy may be able to substitute classic oxidation indices in the determination of oxidative stability of edible oils to its simplicity, low cost, rapidity, and time saving.

During oxidation, there are some changes in their IR spectra of edible oils caused by changes in their compositions, especially in the regions around 3500, 1700, 1300–1200, 1050–800 and 575 cm^{-1} as reported by Moros et al. [116]. The authors investigated the FTIR spectra of un-oxidized sunflower oil and heated sunflower oil at 147 °C during oxidation for 1920 min.

Vlackos et al. [105] investigated the oxidative stability of corn oil during frying using FTIR spectroscopy. The samples were subjected to heating up to various temperatures from 130 up to 275 °C. The changes in term of peak intensities at frequency region between 3050 and 2740 cm^{-1} take place during the oxidation process. The band intensities at 2854 cm^{-1} and the shoulder at 2962 cm^{-1} were increased, meanwhile the band intensity at 2925 cm^{-1} was reduced as the temperature gets higher. A shoulder at 2872 cm^{-1} is also formed which corresponds to the production of methyl (CH_3) groups. Moya-Moreno et al. [119] has monitored the thermal oxidation of fatty acids in edible oils using FTIR spectroscopy at 80 – 300 °C, for 20 – 40 min. The results revealed that there is a decrease in unsaturation observed starting at 150°C and becoming more pronounced at around 250°C. Furthermore, the same authors [120] also had investigated the combination of FTIR and H-NMR spectroscopies in identifying all the oxidation products. The calibration set was prepared by adding butyraldehyde with known concentrations, which represented aldehydic and ketonic compounds produced during oxidation process. An exponential increase of carbonylic compounds with temperature was also observed after heating at 150°C. From the results, the authors conclude that the combination of FTIR and H-NMR spectroscopies is excellent technique to provide information on composition and chemical nature of the heated oils.

The oxidative stability of edible oils was also investigated with FTIR spectroscopy approach using disposable polymer IR cards [121]. To facilitate the oxidation process, pure triacyglycerol-triolein, trilinolein, and trilinolenin were loaded onto the cards and located in chamber at 55°C. Using peak-find function, the minimum absorbance in the *cis* region (3017-3000 cm^{-1}) and the maximum absorbance of hydroperoxide functional group (3550-3200cm^{-1}), isolated *trans* (977-957 cm^{-1}), and conjugated *trans* regions (995-983 cm^{-1}) were measured at the differential spectra and plotted as a function of time.

In conclusion, the development in IR spectroscopy instrumentation, particularly from dispersive to FTIR spectrometer combined with powerful chemometric tools have increased the performance of FTIR spectroscopy especially in analysis of fats and oils. The advent of FTIR spectroscopy has revitalized quantitative IR spectroscopy owing to the superior performance and advanced data handling capabilities of FTIR systems. In recent years, a substantial amount of researches has been focused on the possible applications of IR spectroscopy in the fats and oils industry, as well as to the food industry. This interest has been driven by increased demand in the industry for automated instrumental methods that can replace the traditional wet chemical method of analysis. The improvements in efficiency and analytical performance of FTIR spectroscopy can provide an important aspect, probably due to the use of solvents and hazardous reagents can be avoided.

The development of rapid spectroscopic methods to evaluate fats and oils quality is desirable, not only to monitor their quality but also to verify the authenticity of the products as well as to monitor oxidative stability of fats and oils. FTIR spectroscopy has shown its potential application in determining and monitoring various fats and oils quality. FTIR spectroscopy also benefits the food industry by providing rapid methods for detecting adulteration and confirming authenticity of the food. Food authentication is important to ensure the food safety and protecting consumers from fraud and deception as well as for product recall purposes. It is also widely used for monitoring of fats and oils deterioration due to oxidation process.The use of FTIR spectroscopy fulfilled the need for fast, safe and reliable instrumental methods for testing physical and chemical properties of fats and oils and other related products.

ACKNOWLEDGMENTS

This study was supported by University Putra Malaysia through Research University Scheme Grant (RUGS) with No. 02/01/07/0032RU.

REFERENCES

[1]   Roggo, Y.; Chalus, P.; Maurer, L.; Lema-Martinez, C.; Edmond, A.; Jent, N. A Review of near infrared spectroscopy and chemometrics in pharmaceutical technologies. *J. Pharm. Biomed. Anal.* 2007, *44*, 683 – 700.

[2]   Wang,W.; Paliwal, J. Near-infrared spectroscopy and imaging in food quality and safety. *Sens. Instrument. Food Qual..* 2007, *1*, 193–207.

[3]   Sun, D-W., Ed. Modern Techniques for food authentication; 1st Ed.; Elsevier: London., 2008; Chapter 2. pp 27 – 57.

[4]   Pare J.R.J.; Belanger Ismail, J.M.R., Ed. Food Analysis, Elsevier: London., 1997; Chapter 4. pp. 93-139

[5]   Downey, G. Food and food ingredient authentication by mid-infrared spectroscopy and chemometrics. *Trends Anal. Chem.* 1998. *17*, 418 -424.

[6]   Sun, D-W., Ed. Infrared Spectroscopy for Food Quality Analysis and Control; 1st Ed. Elsevier: London 2009; Chapter 1. pp 1 -25.

[7]   Stuart, B. *Modern Infrared Spectroscopy*, edited by Ando, D.J., John Wiley and Sons: New York, 1996; pp 135.

[8]   Pavia, D.L.; Lampman, G.M.; Kriz-jr, G.S. *Introdoction to Spectroscopy: A Guide for Students of Organic Chemistry*, 3th Ed.; Thomson Learning Inc: London, 2001. pp 14-24.

[9]   Fredericks, P.M.; Lee, J.B.; Osborn, P.R.; Swinkels, D.A.J. Materials Characterizations Using Factor Analysis of Fourier Transform Infrared Spectra, Part 1: Results. *J. Appl. Spectros,* 1985, 39, 303-310.

[10] Cast, J. Infrared Spectroscopy of lipids in *Development in Oils and Fats*, Blackie Academic and Professional: London, 1992; pp. 224 – 266.

[11] Wilson, R.H.; Godfellow, B.J. Mid-Infrared Spectroscopy in *Spectroscopic Techniques for Food Analysis*, edited by Wilson, R.H., VCH Publishers, Inc.: New York, 1994. pp 59 – 84.

[12] Siverstein, R.M.; Webster, F.X. *Spectrometric identification of organic compouns*, 6th Edition John Wiley and Sons: New York, 1998; pp. 71 – 78.

[13] Guillen, M.D.; Cabo, N. Infrared Spectroscopy in the Study of Edible Oils and Fats. *J Sci. Food Agric.* 1997, *75*, 1-11.

[14] Rohman, A.; Che Man, Y. B. Fourier transform infrared (FTIR) spectroscopy for analysis of extra virgin olive oil adulterated with palm oil. *Food Res. Int.* 2010, *43*, 886 – 892.

[15] Guillen, M.D.; Cabo, N. Characterization of Edible Oils and Lard by Fourier Transform Infrared Spectroscopy: Relationships Between Composition and Frequency of Concrete Bands in the Fingerprint Region. *J. Am. Oil Chem. Soc,* 1997, *74*, 1281 – 1286.

[16] Perkins, W.D. Sample handling in infrared spectroscopy- an overview, in Practical sampling techniques for infrared analysis, edited by Patricia Coleman, CRC Press Inc: London, 1993; pp 1 – 53.

[17] Wilson, R.H.; Tapp, H.S. Mid-infrared spectroscopy for food analysis: recent new applications and relevant developments in sample presentation methods. *Trends Anal. Chem.* 1999, *18*, 85 – 93.

[18] Ismail, A.A.; Nicodemo, A.; Sedman, J.; van de Voort, F.R.; Holzbaur, E. Infrared Spectroscopy of Lipid: Principles and Applications in *Spectral Properties of Lipids*, Sheffield Academic Press Ltd.: England, 1999; pp 235-269.

[19] Smith, A.L. *Applied Infrared Spectroscopy: Fundamentals, Techniques and Analytical Problem Solving,* John Wiley and Sons: New York, 1979.pp 74 -83.

[20] American Society for Testing Materials. Standard Practice for Internal Reflection Spectroscopy, Standard Practice E573-90, American Society for Testing Materials, Philadelphia, PA. In *Practical Sampling Techniques for Infrared Analysis*, edited by P.B. Coleman, CRC Press: USA, 1990; pp. 56-63.

[21] Sedman, J.; van de Voort, F. R.; Ismail, A. A. Attenuated Total Reflectance Spectroscopy: Principles and Applications in Infrared Analysis of Food in Spectral Methods in Food Analysis, edited by Mossoba, M.M., Marcel and Decker Inc: New York, 1999.pp 397 – 427.

[22] Perkin Elmer. FT-IR Spectroscopy Attenuated Total Reflectance (ATR). https://las.perkinelmer.com/content/TechnicalInfo/TCH_FTIRATR.pdf.

[23] Compton S.V.; Compton, D.A.C. Quantitative Analysis-Avoiding Common Pitfalls, in *Practical Sampling Techniques for Infrared Analysis*, edited by Coleman, P.B., CRC Press: UK, 1993; pp 227-230.

[24] Lavine, B.K. Chemometrics. *Anal. Chem.* 1998, *70*, 209R-228R.

[25] Rutan, S.C. Discovering chemistry with chemometrics. *Chemomet. Intell. Lab. Sys.* 1992. *15*, 137-141.

[26] Aparicio, R.; Aparicio-Ruiz, R. (2002). Chemometrics as an Aid in Authentication, *in Oils and Fats Authentication*, edited by Jee, M. Blackwell Publishing Ltd.: UK, 2002; pp 156-180.

[27] Lai, Y.W.; Kemsley, E.K.; Wilson, R.H. Potential of Fourier Transform Infrared Spectroscopy for the Authentication of Vegetable Oils. *J. Agric. Food Chem.* 1994, *42*, 1154-1159.

[28] Arvanitoyannis, I.S.; Vaitsi, O.B. A Review on Tomato Authenticity: Quality Control Methods in Conjunction with Multivariate Analysis (Chemometrics). *Crit. Rev. Food Sci. Nutr.* 2007, *47*, 675 – 699.

[29] Arvanitoyannis, I.S. and Tzouros, N.E Implementation of Quality Control Methods in Conjunction with Chemometrics Toward Authentication of Dairy Products. *Crit. Rev. Food Sci. Nutr.* 2005, *45,* 231-249.

[30] Arvanitoyannis, I.S.; van Houwelingen-Koukaliaroglou, M. Implementation of Chemometrics for Quality Control and Authentication of Meat and Meat Products. *Crit. Rev. Food Sci. Nutr.* 2003, *43*, 173–218.

[31] de la Fuente, M.A., and Juarez, M. Authenticity Assessment of Dairy Products. *Crit. Rev. Food Sci. Nutr.* 2005, *45*, 563–585.

[32] Maggio, R.M.; Kaufman, T.S.; Del Carlo, M.; Cerretani, L.; Bendini, A.; Cichelli, A.; Compagnone, D. Monitoring of fatty acid composition in virgin olive oil by Fourier transformed infrared spectroscopy coupled with partial least squares. *Food Chem.* 2009, *114,* 1549–1554.

[33] Paradkar, M.M.; Sivakesava, S.; Irudayaraj, J. Discrimination and classification of adulterants in maple syrup with the use of infrared spectroscopic techniques. *J. Sci. Food Technol.* 2002, *82,* 497 – 504.

[34] Che Man, Y.B.; Syahariza, Z.A.; Mirghani, M.E.S.; Jinap, S.; Bakar J. Analysis of potential lard adulteration in chocolate and chocolate products using Fourier transform infrared spectroscopy. *Food Chem.* 2005, *90,* 815–819.

[35] Smith, B.C. *Quantitative spectroscopy: Theory and Practice*, 1st Ed., Academic Press: Amsterdam, 2002; pp. 125 – 179.

[36] Ballabio, D.; Todeschini, R. *Multivariate classification for qualitative analysis in infrared spectroscopy for food quality: analysis and control* edited by Da-Wen Sun, Elsevier: New York, 2009, pp. 83 – 104.

[37] Manaf, M.A.; Che Man, Y.B.; Hamid, N.S.A.; Ismail, A.; Syahariza, Z.A. Analysis of adulteration of virgin coconut oil by palm kernel olein using Fourier transform Infrared spectroscopy. *J. Food Lipids.* 2007, *14,* 111–121.

[38] Rohman, A.; Che Man, Y.B. Analysis of Cod-Liver Oil Adulteration Using Fourier Transform Infrared (FTIR) Spectroscopy. *J. Am. Oil Chem. Soc.* 2009, *86,* 1149-1153.

[39] Van de Voort, F.R.; Ghetler, A.; García-González, D.; Li, Y. Perspectives on Quantitative Mid-FTIR Spectroscopy in Relation to Edible Oil and Lubricant Analysis: Evolution and Integration of Analytical Methodologies. *Food Anal. Methods.* 2008, *1,* 153-163.

[40] Rossell, J.B. Classical Analysis of Oils and Fats, in *Analysis of Oils and Fats*, edited by Hamilton, R.J. and Rossell, J.B. Elsevier Applied Science: England, 1986; pp 1-90.

[41] Miyake, Y.; Yokomizo, K.; Matsuzaki, N. Rapid Determination of Iodine Value by Nuclear Magnetic Resonance Spectroscopy. *J. Am. Oil Chem. Soc.* 1998, *75,* 15-19.

[42] AOCS (1992). Official Methods and Recommended Practices of The American Oil Chemists' Society, 4th Edition, Method Cd 1-25, Champaign II: AOCS Press.

[43] ISO (1989) *Animal and Vegetable Fats and Oils*- Determination of Iodine Value, Method ISO 3961, International Organization for Standardization, Swisszerland.

[44] Hui, Y.H. *Bailey's Industrial Oil and Fat Products*, Volume 1, 5th Ed., John Wiley and Sons, Inc: New York, 1996.

[45] van de Voort, F. R.; Sedman, J.; Emo, G.; Ismail, A.A. Rapid and Direct Iodine Value and Saponification Number Determination of Fats and Oils by Attenuated Total Reflectance/Fourier Transform Infrared Spectroscopy. *J. Am. Oil Chem. Soc.* 1992, *69,* 1118 – 1123.

[46] Kampars, V.; Ronberga, S. Iodine values estimation of vegetable oils by FTIR spectroscopy. *Polish J. Food Nutr. Sci.* 2003, *12/53,* 45-47.

[47] Hendl, O.; Howell, J.A.; Lowery, J.; Jones, W. A rapid and simple method for the determination of iodine values using derivative Fourier transform infrared measurements. *Anal. Chim. Acta.* 2001, *427*, 75–81.

[48] Sedman, J.; van de Voort, F.R.; Ismail, A.A. Simultaneous Determination of Iodine Value and trans Content of Fats and Oils by Single-Bounce Horizontal Attenuated Total Reflectance Fourier Transform Infrared Spectroscopy. *J. Am. Oil Chem. Soc.* 2000, *77*, 399 – 403.

[49] Che Man Y.B.; Setiowaty, G. Multivariate Calibration of Fourier transform infrared spectra in determining iodine value of palm oil products. *Food Chem.* 1999, *67*, 193-198.

[50] Che Man Y.B., Setiowaty, G., and van de Voort, F.R. Determination of Iodine Value of Palm Oil by Fourier Transform Infrared Spectroscopy. *J. Am. Oil Chem. Soc.* 1999, *76*, 693-700.

[51] Sedman, J.; van de Voort, F.R.; Ismail, A.A.; Maes, P. Industrial Validation of Fourier Transform Infrared trans and Iodine Value Analyses of Fats And Oils. *J. Am. Oil Chem. Soc.* 1998, *75*, 33 – 39.

[52] van de Voort, F. R.; Sedman, J.; Emo, G. A Rapid FTIR quality Control Method for Fat and Moisture Determination in Butter. *Food Res. Int.* 1992, *25,*193-198.

[53] Che Man, Y.B.; Setiowaty, G. Application of Fourier transform infrared spectroscopy to determine free fatty acid contents in palm olein. *Food Chem.* 1999, *66*, 109 – 114.

[54] Che Man Y.B.; Moh, M.H.; van de Voort, F.R. Determination of Free Fatty Acids in Crude Palm Oil and Refined-Bleached-Deodorized Palm Olein Using Fourier Transform Infrared Spectroscopy. *J. Am. Oil Chem. Soc.* 1999, *76*, 485-490.

[55] Bertran, E.; Blanco, M.; Coello, J.; Iturriaga, H.; Maspoch, S.; Montoliu, I. Determination of Olive Oil Free Fatty Acid by Fourier Transform Infrared Spectroscopy. *J. Am. Oil Chem. Soc.* 1999, *76*, 611-616.

[56] Sherazi, S. T. H.; Mahesar, S. A.; Bhanger, M. I.; van de Voort, F. R.; Sedman, J. Rapid Determination of Free Fatty Acids in Poultry Feed Lipid Extracts by SB-ATR FTIR Spectroscopy. *Journal of Agriculture and Food Chem.* 2007, *55*, 4928–4932.

[57] Guillen M.D; Cabo, N. Fourier transform infrared spectra data versus peroxide and anisidine values to determine oxidative stability of edible oils. *Food Chem.* 2002, *77*, 503-510.

[58] Ruíz, A.; Cañada, M.J. A.; Lendl, B. A rapid method for peroxide value determination in edible oils based on flow analysis with Fourier transform infrared spectroscopic detection. *Analyst.* 2001, *126*, 242–246.

[59] Setiowaty G.; Che Man, Y.B.; Jinap, S.; Moh, M.H. Quantitative Determination of Peroxide Value in Thermally Oxidized Palm Olein by Fourier Transform Infrared Spectroscopy. *Phytochem. Anal.* 2000, *11*, 74-78.

[60] Moh, M.H.; Tang, T.S.; Che Man, Y.B.; Lai, O.M. Rapid Determination of Peroxide Value in Crude Palm Oil Products Using Fourier Transform Infrared Spectroscopy. *J. Food Lipids.* 1999, *6*, 261-270.

[61] Ma, K.; van de Voort, F.R.; Sedman, J.; Ismail, A.A. Stoichiometric Determination of Hydroperoxide in Fats and Oils by Fourier Transform Infrared Spectroscopy. *J. Am. Oil Chem. Soc.* 1997, *74*, 897-906.

[62] van de Voort, F.R.; Ismail, A.A.; Sedman, J.; Emo, G. Monitoring the Oxidation of Edible Oils by Fourier Transform Infrared Spectroscopy. *J. Am. Oil Chem. Soc.* 1994, *71*, 243-253.

[63] Che Man Y.B; Setiowaty, G. Determination of Anisidine Value in Thermally Oxidized Palm Olein by Fourier Transform Infrared Spectroscopy. *J. Am. Oil Chem. Soc.* 1999, *76*, 243-247.

[64] Godoy, S.C.; Ferra~o, M.F.; Gerbase, A.E. Determination of the Hydroxyl Value of Soybean Polyol by Attenuated Total Reflectance/Fourier Transform Infrared Spectroscopy. *J. Am. Oil Chem. Soc.* 2007, *84*, 503–508.

[65] Setiowaty, G.; Che Man, Y.B. A rapid Fourier transform infrared spectroscopic method for the determination of 2-TBARS in palm olein. *Food Chem.* 2003, *81*, 147–154.

[66] Mirghani, M.E.S.; Che Man, Y.B.; Jinap, S.; Baharin, B.S.; Bakar, J. Rapid Method for Determining of Malonaldehyde as Secondary Oxidation Product in Palm Olein System by FTIR Spectroscopy. *Phytochem. Anal.* 2002, *3(4)*, 195-201.

[67] Mirghani, M.E.S.; Che Man, Y.B.; Jinap, S.; Baharin, B.S.; Bakar, J. Multivariate Calibration of Fourier Transform Spectra in Determining the Malonaldehyde as a Thiobarbituric Acid Reactive Substances (TBARS) in Palm Olein. *J. Am. Oil Chem. Soc.* 2001, *78*, 1127-1131.

[68] Capote, F.P.; Jiminez, J.R.; Olmo, J.G.; de Castro, M.D. Fast Method for the Determination of Total Fat and *trans* Fatty Acids Content in Bakery Products Based on Microwave-Assisted Soxhlet Extraction and Medium Infrared Spectroscopy Detection. *Anal. Chim. Acta.* 2004, *517*, 13-20.

[69] Mossoba, M.M.; Yurawecz, M.P.; McDonald, R.E. Rapid Determination of the Total *trans* Content of Neat Hydrogenated Oils by Attenuated Total Reflection Spectroscopy. *J. Am. Oil Chem. Soc.* 1996, *73*, 1003-1009.

[70] Dubois, J.; van de Voort, F.R.; Sedman, J.; Ismail, A.A.; Ramaswamy, H.R. Quantitative Fourier Transform Infrared Analysis for Anisidine Value and Aldehydes in Thermally Stressed Oils. *J. Am. Oil Chem. Soc.* 1996, *73*, 787-794.

[71] Ulberth, F.; Haider, H.J. Determination of Low Level Trans Unsaturation in Fats by Fourier Transform Infrared Spectroscopy. *J. Food Sci.* 1992, *57*, 1444-1447.

[72] Lancer, A.C.; Emken, E.A. Comparison of FTIR and Capillary Gas Chromatographic Methods for Quantification of trans unsaturation in Fatty Acid Methyl Esters. *J. Am. Oil Chem. Soc.* 1988, *68*, 448-449.

[73] Madison, B.L.; De Palma, R.A.; D'Alonzo, R.P. Accurate Determination of Trans Isomers in Shortening and Edible Oils by Infrared Spectrophotometry. *J. Am. Oil Chem. Soc.* 1982, *59*, 178-181.

[74] Setiowaty, G. (2003). *FTIR as Tools in Conjunction with Chemometric Methods Applied to Palm Oil Analysis*, (Ph.D Disertation), Universiti Putra Malaysia, Selangor, Malaysia.

[75] Che Man, Y.B.; Mirghani, M.E.S. Rapid Determinations of Moisture Content in Crude Palm Oil by Fourier Transform Infrared Spectroscopy. *J. Am. Oil Chem. Soc.* 2000, *77*, 631-637.

[76] Setiowaty G.; Che Man, Y.B. Determination of Slip Melting Point in Palm Oil Blends by PLS and PCR Modeling of FTIR Spectroscopic Data. *J. Am. Oil Chem. Soc.* 2002, *79*, 1081-1084.

[77] Setiowaty, G.; Che Man, Y.B. Multivariate Determination of Cloud Point in Palm Oil using Partial Least Squares and Principal Component Regression Based on FTIR Spectroscopy. *J. Am. Oil Chem. Soc.* 2004, *81*, 7-11.

[78] Yang, H.; Irudayaraj, J. Characterization of Semisolid Fats and Edible Oils by Fourier Transform Infrared Photoacoustic Spectroscopy. *J. Am. Oil Chem. Soc.* 2000, *77*, 291-295.

[79] Inon, F.A.; Garrigues J.M.; Garrigues, S.; Molina, A.; de la Guardia, M. Selection of Calibration Set samples in Determination of Olive Oil Acidity by Partial Least Squares-Attenuated Total Reflectance-Fourier Transform Infrared Spectroscopy. *Anal. Chim. Acta.* 2003, *89*, 59-75.

[80] Ismail, A.A.; van de Voort, F.R.; Emo, G.; Sedman, J. Rapid and Quantitative Determination of Free Fatty Acids by Fourier Transform Infrared Spectroscopy. *J. Am. Oil Chem. Soc.* 1993, *70*, 335-341.

[81] Li, Y.; Garcı́a-González, D.L.; Yu, X.; van de Voor, F.R. Determination of Free Fatty Acids in Edible Oils with the Use of a Variable Filter Array IR Spectrometer. *J. Am. Oil Chem. Soc.* 2008, *85*, 599–604.

[82] Yu, X.; Du, S.; van de Voort, F.R.; Yue, T.; Li, Z. Automated and simultaneous determination of free fatty acids and peroxide values in edible oils by FTIR spectroscopy using spectral reconstitution. *Anal. Sci.* 2009, *25*, 627-632.

[83] Aryee, A.N.A.; van de Voort, F.R.; Simpson, B.K. FTIR determination of free fatty acids in fish oils intended for biodiesel production. *Process Biochem.* 2009, *44*, 401–405.

[84] Verleyen, T.; Verhe, R.; Huyghebaert, A.; Greyt, W.D. Influence of Triacylglycerol Characteristics on the Determination of Free Fatty Acids in Vegetable Oils by Fourier Transform Infrared Spectroscopy. *J. Am. Oil Chem. Soc.* 2001, *78*, 981-984.

[85] Engelson, S.B. Explorative Spectroscopic Evaluations on Frying Oil Deterioration. *J. Am. Oil Chem. Soc.* 1997, *74*, 1495-1508.

[86] Paquette, G.; Kupranycz, D.B.; van de Voort, F.R. The Mechanisms of Lipid Oxidation 1. Primary Oxidation Products. *J. Can. Inst. Food Sci. Technol.* 1985, *18*, 112-118.

[87] Pokorny, J.; Yanishlieva, N.; Gordon, M. Antioxidant in Food: Practical Applications. CRC Press: New York, 2001.

[88] Mensink, R.P.; Katan, M.BEffect of Dietary *trans* Fatty Acids on High-Density and Low-Density Lipoprotein Cholesterol Levels in Healthy Subject. *New Engl. J. Med.* 1990, *323*, 439.

[89] Sleeter, R.I.; Matlock, M.G. Automated Quantitative Analysis of Isolated (nonconjugated) trans Isomers using FTIR Spectroscopy Incorporating Improvements in the Procedure. *Anal. Chem.* 1989, *66*, 121-127.

[90] Sedman, J.; van de Voort, F. R.; Ismail, A. A. Upgrading the AOCS Infrared *trans* Method for Analysis of Neats Fats and Oils by Fourier Transform Infrared Spectroscopy. *J. Am. Oil Chem. Soc.* 1997, *74*, 907-913.

[91] Moh, M.H.; Che Man, Y.B.; Baharin, B.S.; Jinap, S.; Saad, M.S.; Abdullah, W.J.W. Quantitative Analysis of Palm Carotene Using Fourier Transform and Near Infrared Spectroscopy. *J. Am. Oil Chem. Soc.* 1999, *76*, 249-254.

[92] Mirghani, M.E.S.; Che Man, Y.B.; Jinap, S.; Baharin, B.S.; Bakar, J. Application of FTIR Spectroscopy in Determining Sesamol in Sesame Seed Oil. *J. Am. Oil Chem. Soc.* 2003, *80*, 1- 4.

[93] Mirghani, M.E.S.; Che Man, Y.B. A New Method for Determining Gossypol in Cottonseed Oil by Fourier Transform Infrared Spectroscopy. *J. Am. Oil Chem. Soc.* 2003, *80,* 625-628.

[94] Mirghani, M.E.S.; Che Man, Y.B.; Jinap, S.; Baharin, B.S.; Bakar, J. A New Method for Determining Aflatoxins in Groundnut and Groundnut Cake using Fourier Transform Infrared Spectroscopy with Attenuated Total Reflectance Technique. *J. Am. Oil Chem. Soc.* 2001, *78,* 985-992.

[95] Che Man, Y.B.; Ammawath, W.; Mirghani, M.E.S. Determining α- tocopherol in Refined Bleached and Deodorized Palm Olein by Fourier Transform Infrared Spectroscopy. *Food Chem.* 2005, *90,* 323–327.

[96] Ammawath, W.; Che Man, Y.B.; Baharin, B.S.; Abdul Rahman, R.B. Multivariate determination of propyl gallate in RBD palm olein using partial least squares and principal component regression based on FTIR spectroscopy. *J. Food Lipids.* 2006, *13,* 1-11.

[97] Ammawath, W.; Che Man, Y.B.; Baharin, B.S.; Abdul Rahman, R.B. Analysis of butylated hydroxyanisole (BHA) in RBD palm oil and RBD palm olein using partial least squares based on FTIR spectroscopy. *J. Food Lipids.* 2005, *12,* 198-208.

[98] Ammawath, W.; Che Man, Y.B.; Baharin, B.S.; Abdul Rahman, R.B. A new method for determination of *tert*-butylhydroquinone (TBHQ) in RBD palm olein with FTIR spectroscopy. *J. Food Lipids.* 2005, *11,* 266–277.

[99] Reid, L.M.; O'Donnell, C.P.; Downey, G. Recent technological advances for the determination of food authenticity. *Trends Food Sci. Technol.* 2006, *17,* 344–353.

[100] Rossell, J. B.; King, B.; Downes, M. J. Detection of adulteration. *J. Am. Oil Chem. Soc.* 1983, *60,* 333–339.

[101] Rohman, A.; Che Man, Y.B. Monitoring of virgin coconut oil (VCO) adulteration with palm oil using Fourier transform infrared spectroscopy. J. *Food Lipids.* 2009, *16,* 618–628.

[102] Lerma-García, M.J.; Ramis-Ramos, G.; Herrero-Martínez, J.M.; Simó-Alfonso, E.F. Authentication of extra virgin olive oils by Fourier-transform infrared spectroscopy. *Food Chem.* 2010, *118,* 78–83.

[103] Gurdeniz, G.; Ozen, B. Detection of adulteration of extra-virgin olive oil by chemometric analysis of mid-infrared spectral data. *Food Chem.* 2009, *116,* 519–525.

[104] Alam, M.A.; Hamid, S.F. Application of FTIR spectroscopy in the assessment of olive oil adulteration. *J. Appl. Sci. Res.* 2007, *73,* 102–108.

[105] Vlachos, N.; Skopelitis, Y.; Psaroudaki, M.; Konstantinidou, V.; Chatzilazarou, A.; Tegou, E. Applications of Fourier transform-infrared spectroscopy to edible oils. *Anal. Chim. Acta.* 2006, *573–574,* 459–465.

[106] Downey, G.; Mc Intyre, P.; Davies, A. N. Detecting and Quantifying Sunflower Oil Adulteration in Extra Virgin Olive Oils from the Eastern Mediterranean by Visible and Near-Infrared Spectroscopy. *J. Agric.Food Chem.* 2002, *50,* 5520☐-5525.

[107] Yang, H.; Irudayaraj, J. Comparison of Near-Infrared, Fourier Transform-Infrared, and Fourier Transform-Raman Methods for Determining Olive Pomace Oil Adulteration in Extra Virgin Olive Oil. *J. Am. Oil Chem. Soc.* 2001, *78,* 889–895.

[108] Marigheto, N.A.; Kemsley, E.K.; Defernez, M.; Wilson R.H. A Comparison of mid-infrared and raman spectroscopies for the authentication of edible oils. *J. Am. Oil Chem. Soc.* 1998, *75,* 987–992.

[109] Wesley, I.l.; Pacheco, E.; McGill, A.E.J. Identification of adulterants in olive oils. *J. Am. Oil Chem. Soc.* 1996, *73,* 515–518.

[110] Christy, A.A.; Kasemsumran, S.; Du, Y.; Ozaki, K. The detection and quantification of adulteration in olive oil by near-infrared spectroscopy and chemometrics. *Anal. Sci.* 2004, *20,* 935-940.

[111] Tay, A.; Singh, R. K.; Krishnan S. S.; Gore J. P. Authentication of Olive Oil Adulterated with Vegetable Oils Using Fourier Transform Infrared Spectroscopy, *Lebensm.-Wiss. u.-Technol.* 2002, *35,* 99–103.

[112] Wang, L.; Lee; F.S.C.; Wang, X.; He, Y. Feasibility study of quantifying and discriminating soybean oil adulteration in camellia oils by attenuated total reflectance MIR and fiber optic diffuse reflectance NIR. *Food Chem.* 2006, *95,* 529–536.

[113] Baeten, V.; Pierna, J.A.F.; Dardenne, P.; Meurens, M.; García-González D.L.; Aparicio-Ruiz, R. Detection of the presence of hazelnut oil in olive oil by FT-Raman and FT-MIR spectroscopy. *J. Agric.Food Chem.* 2005, *53,* 6201–6206.

[114] Syahariza, Z.A.; Che Man, Y.B.; Selamat, J.; Bakar J. Detection of lard adulteration in cake formulation by Fourier transform infrared (FTIR) spectroscopy. *Food Chem.* 2005, *92,* 365–371.

[115] Lai, Y.W.; Kemsley, E.K.; Wilson, R.H. Quantitative Analysis of Potential Adulterants of Extra Virgin Olive Oil using Infrared Spectroscopy. *Food Chem.* 1995, *53,* 95-98.

[116] Moros, J.; Roth, M.; Garrigues, S.; de la Guardia, M. Preliminary studies about thermal degradation of edible oils through attenuated total reflectance mid-infrared spectrometry. *Food Chem.* 2009, *114,* 1529–1536.

[117] Muik, B.; Lendl, B.; Molina-Dıaz, A.; Ayora-Ca˜nada, M.J. Direct monitoring of lipid oxidation in edible oils by Fourier transform Raman spectroscopy. *Chem. Phys. Lipids.* 2005, *134,* 173–182.

[118] van de Voort, F. R. FTIR Spectroscopy in edible oil analysis. *INFORM.* 1994, *5,* 1038 - 1042.

[119] Moreno, M.M.C.M.; Olivares, D.M.; Lopez, F.J.A.; Adelantado, J.V.G.; Reig, F.B. Analytical evaluation of polyunsaturated fatty acids degradation during thermal oxidation of edible oils by Fourier transform infrared spectroscopy. *Talanta.* 1999, *50,* 269–275.

[120] Moreno, M.M.C.M.; Olivares, D.M.; Lopez, F.J.A.; Adelantado, J.V.G.; Reig, F.B. Determination of unsaturation grade and trans isomers generated during thermal oxidation of edible oils and fats by FTIR. *J. Mol. Struct.* 1999, *482-483,* 551-556.

[121] Russin, T.A.; van de Voort, F.R.; Sedman, J. Novel method for rapid monitoring of lipid oxidation by FTIR Spectroscopy using disposable IR cards. *J. Am. Oil Chem. Soc.* 2003, *80,* 635-641.

In: Fourier Transform Infrared Spectroscopy
Editor: Oliver J. Rees, pp. 27-78

ISBN: 978-1-61668-835-6
© 2010 Nova Science Publishers, Inc.

Chapter 2

A MODIFIED FOURIER SERIES METHOD FOR THE DYNAMIC ANALYSIS OF STRUCTURES

Wen L. Li and *Hongan Xu*

Department of Mechanical Engineering, Wayne State University
5050 Anthony Wayne Drive, Detroit, Michigan 48202, U.S.A

ABSTRACT

A modified Fourier series method is described for solving various structural dynamic problems. To better understand the essence of this new approach, a brief review is first given of the conventional Fourier series expansion and a few related mathematical theorems. An improved Fourier series representation is then present which can be used to expand any function, over a solution domain including the boundary points, with a pre-determined rate of convergence. Thus, for a given boundary value problem, an exact continuous solution can be systematically obtained by letting the series simultaneously satisfy both the governing differential equations and the boundary conditions on a point-wise basis. This improved Fourier series method is first used to determine the vibrations of beams with general boundary conditions. It is subsequently extended to the vibrations of two arbitrarily coupled beams, multi-span beams under moving loads, plates with arbitrary boundary supports, and built-up structures composed of any number of beams and plates. The excellent accuracy and convergence of the analytical solutions have been repeatedly demonstrated by numerical examples with varying degrees of difficulties. The improved Fourier series method actually represents a general and powerful mathematical technique for solving a wide range of boundary value problems including the vibrations of various structural components and systems.

INTRODUCTION

The vibrations of beams, plates and shells have been extensively studied in structural dynamics. Since a complex dynamic system can be viewed as an assembly of a number of

---

* Email: wli@wayne.edu

basic structural components in engineering analysis and design, the characteristic for each individual member will understandably affect the system behavior to some extent. Therefore, not only is the study of these basic structural members important to the design of complex systems, but also the component-level solution techniques are usually applicable to system-level modeling. For example, the beam functions (the solutions of simple beam problems) are often used to expand the transverse displacements in plate problems [1], or construct the general solutions for complex structures as in Spectral Element Methods (SEM) [2-4].

A wide spectrum of techniques has been developed for the dynamic analysis of beams under various load or boundary conditions. Among them, the modal superposition technique is probably the most popular one in which the beam displacement is expressed as a linear combination of eigenfunctions or mode shapes. Although eigenfunctions generally exist in the forms of trigonometric and hyperbolic functions, they also include some integration and frequency constants that depend upon boundary conditions. Consequently, each boundary condition essentially calls for a particular set of natural frequencies and mode shapes. However, even consider the simplest end conditions (i.e., pinned, clamped, free and sliding), they can altogether make up 10 different boundary conditions for a beam and 55 different boundary conditions for a (rectangular) plate. Therefore, the use of mode shapes as the basis functions can become very tedious in reality. Although the frequency equation generally exists for beams with arbitrary elastic supports [5], most investigations in the literature have been primarily focused on some degenerate cases in which the rotational and/or translational springs are arranged in certain special ways [6-8].

The beam displacement can also be sought in forms of, such as, polynomials [9, 10], Fourier series [11] or other functions [12, 13]. Trigonometric functions are perhaps the most desired set in expanding a function or an analytical solution due to their completeness, orthogonality, and excellent numerical stability. Fourier series is widely used to determine the vibrations of simply supported beams. However, a conventional Fourier series representation tends to become slow converged, if at all, for other boundary conditions. In what follows, a modified Fourier series method will be described for solving various structural dynamic problems.

A MODIFIED FOURIER SERIES METHOD FOR SOLVING BOUNDARY VALUE PROBLEMS

It is well known that a continuous function $f(x)$ defined over $[-L, L]$ can always be expanded into a Fourier series as

$$f(x) = a_0 + \sum_{m=1}^{\infty} a_m \cos \lambda_m x + b_m \sin \lambda_m x, \qquad -L < x < L,$$

$$(\lambda_m = \frac{m\pi}{L}) \tag{1}$$

where

$$a_m = \frac{1}{L}\int_{-L}^{L} f(x)\cos\lambda_m x \, dx,$$

$$b_m = \frac{1}{L}\int_{-L}^{L} f(x)\sin\lambda_m x \, dx,$$

and

$$a_0 = \frac{1}{2L}\int_{-L}^{L} f(x)dx.$$

If $f(x)$ is an odd function in the interval, then Eq. (1) reduces to

$$f(x) = \sum_{m=1}^{\infty} b_m \sin\lambda_m x, \qquad\qquad -L < x < L. \tag{2}$$

If $f(x)$ is an even function, then Eq. (1) becomes

$$f(x) = a_0 + \sum_{m=1}^{\infty} a_m \cos\lambda_m x, \qquad\qquad -L < x < L. \tag{3}$$

When certain continuity conditions are satisfied by function $f(x)$, some of its derivatives can be simply obtained through term-by-term differentiation of the series expansions. To better understand this, let's first review a few related mathematical theorems [14].

Theorem 1. *Let $f(x)$ be a continuous function defined on [0, L] with an absolutely integrable derivative, and let $f(x)$ be expanded in Fourier sine series*

$$f(x) = \sum_{m=1}^{\infty} b_m \sin\lambda_m x, \qquad\qquad 0 < x < L,$$

$$(\lambda_m = \frac{m\pi}{L}) \tag{4}$$

then

$$f'(x) = \frac{f(L)-f(0)}{L} + \sum_{m=1}^{\infty}\left(\frac{2}{L}[(-1)^m f(L)-f(0)] + \lambda_m b_m\right)\cos\lambda_m x \cdot \tag{5}$$

Theorem 2. *Let $f(x)$ be a continuous function defined on [0, L] with an absolutely integrable derivative, and let $f(x)$ be expanded in Fourier cosine series*

$$f(x) = a_0 + \sum_{m=1}^{\infty} a_m \cos\lambda_m x, \qquad\qquad 0 < x < L, \qquad\qquad (6)$$

then

$$f'(x) = -\sum_{m=1}^{\infty} \lambda_m a_m \sin\lambda_m x . \qquad\qquad (7)$$

These two theorems basically state that while *a cosine series can always be differentiated term-by-term, this can be done to a sine series if and only if $f(0)=f(L)=0$.* It is important to point out that the two end points of the interval, $x=0$ *and* L, have been explicitly excluded from the Fourier series expansions in Eqs. (4) and (6). In other words, a continuous function can be generally expanded into a Fourier series *inside* the interval (i.e., boundary points are excluded). The word "inside" is highlighted here because it explains why the traditional Fourier method cannot be utilized as a general technique for solving a wide variety of boundary value problems. To illustrate this point, let $w(x)$ be a continuous displacement function for a vibrating beam of length L. *Mathematically, expanding $w(x)$ into a Fourier series simply implies that it is treated as a periodic function (with period L) defined over the entire x axis.*

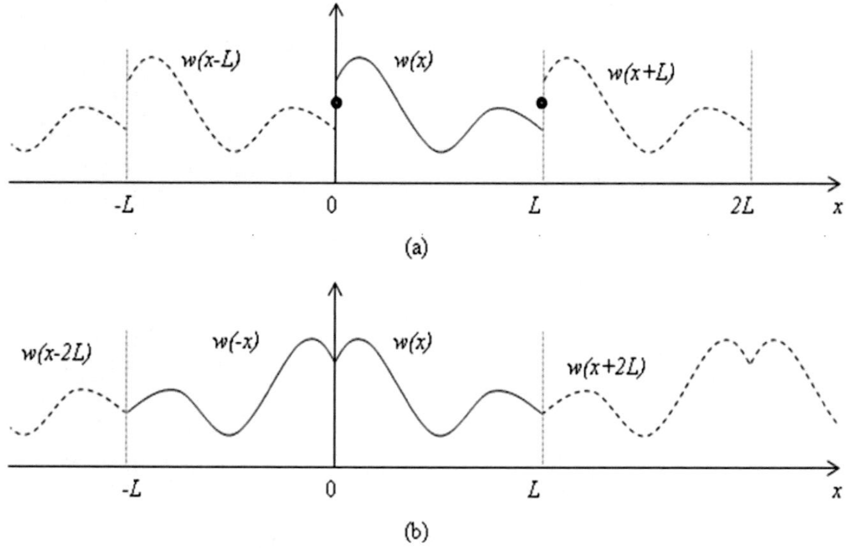

Figure 1. An illustration of the convergence of Fourier series expansions with (a) both cosine and sine terms, and (b) only cosine terms.

Consequently, as shown in Figure 1(a), if the beam is allowed to move at an end (and hence the displacements are most likely to be different at the ends), the Fourier series will only converge to the mean value, $(w(0)+w(L)/2$, at $x=0$, $\pm L$, $\pm 2L$, ... This deems the unsuitability of the conventional Fourier series method since the solution fails to converge to the correct boundary values at $x=0$, and L.

This problem, however, can be easily avoided by viewing the displacement function $w(x)$ as a part of an *even* function of period-2L, as shown in Figure 1(b). The Fourier expansion of this even function will then only contain the cosine terms, and be able to correctly converge to $w(x)$ at *any* point over $[0, L]$. However, this treatment alone cannot revive the Fourier series method because the satisfaction of boundary conditions typically means that, in addition to the values of the displacement, *each of its first three derivatives* has to be correctly produced by the series representation at both ends. Referring to Figure 1(b), the derivative $w'(x)$ clearly represents an odd function over $[-L, L]$. Thus, its (sine) series expansion will invariantly converge to zero at $x=0$ and L, regardless of the actual slopes for the beam. To overcome this problem, we alternatively consider a new function

$$\overline{w}(x) = w(x) - w'(0)\xi_1(x) - w'(L)\xi_2(x) \tag{8}$$

where

$$\xi_1'(x) = \begin{cases} 1, & x = 0 \\ 0, & x = L \end{cases} \tag{9}$$

and

$$\xi_2'(x) = \begin{cases} 0, & x = 0 \\ 1, & x = L \end{cases} \tag{10}$$

Mathematically, $\overline{w}(x)$ represents a residual displacement function which is continuous over $[0, L]$ and has zero-slopes at the both ends. The said conditions can be visualized as if each end of the beam were guided with an elastic restraint against the transverse motion. It is clear that the cosine series representation of $\overline{w}(x)$ is able to converge correctly to the function itself and its derivative at every point on the beam. Since the beam displacement is required to have up to the fourth-order derivatives, two similar terms should also be included in Eq. (8) to remove any possible discontinuities associated with the third-order derivative at the boundary points. What has been said can be mathematically expressed as

$$w(x) = \sum_{m=0}^{\infty} a_m \cos\lambda_m\xi + w'(0)\xi_1(x) + w'(L)\xi_2(x) + w'''(0)\xi_3(x) + w'''(L)\xi_4(x) \tag{11}$$

where

$$\xi_3'''(x) = \begin{cases} 1, & x = 0 \\ 0, & x = L \end{cases} \tag{12}$$

$$\xi_4'''(x) = \begin{cases} 0, & x = 0 \\ 1, & x = L \end{cases} \tag{13}$$

and $\xi_1'''(x) = \xi_2'''(x) = \xi_3'(x) = \xi_4'(x) = 0$ at $x=0$ and L.

The "physical impact" of this mathematical treatment is that an arbitrary boundary condition at each end in the original beam problem is reconfigured into a guided support (that is, zero slope and zero shear force) for the new beam function $\overline{w}(x)$, which is almost automatically sought in the form of a cosine expansion or modal superposition (which may be a preferred term by a dynamist).

Although the beam displacement function is specifically mentioned in the above discussions, the series representation given in Eq. (11) is actually able to expand and uniformly converge to *any* function $f(x) \in C^4$ for $\forall (x) \in [0, L]$. Also, this series can be directly differentiated, term-by-term, to obtain uniformly convergent series expansions for up to the fourth-order derivatives of the function $f(x)$. Mathematically, an exact displacement (or strong form of) solution is a *particular* function $w(x) \in C^3$ which satisfies the governing equation at *every* field point and the boundary conditions at *every* boundary point. Thus, the task of seeking an exact solution is now simply turned into finding a set of expansion coefficients to ensure the governing equation and the boundary conditions to be satisfied by the current series solution *exactly on a point-wise basis*. In the following, this modified Fourier series method will be specifically applied to the vibrations of beams with general boundary conditions.

VIBRATIONS OF BEAMS WITH GENERAL BOUNDARY CONDITIONS

Figure 2 shows a beam with general elastic restraints at each end. The governing differential equation for the free vibration of a beam is

$$D\,d^4w(x)\big/d\,x^4 - \rho A \omega^2 w(x) = 0 \tag{14}$$

or

$$w''''(x) - \rho_D \omega^2 w(x) = 0 \tag{15}$$

where D, ρ and A are, respectively, the flexural rigidity, the mass density and the cross-sectional area of the beam, ω is frequency in radian, and $\rho_D = \rho A / D$.

The boundary conditions can be generally expressed as

$$k_0 w = -Dw''', \qquad K_0 w' = Dw'', \qquad\qquad \text{at } x = 0 \tag{16, 17}$$

and

$$k_1 w = Dw''', \qquad K_1 w' = -Dw'', \qquad\qquad \text{at } x = L \tag{18, 19}$$

where k_0 and k_1 are the linear spring constants, and K_0 and K_1 are the rotational spring constants at $x=0$ and L, respectively. It is clear that Eqs. (16-19) represent a set of general boundary conditions; all the classical (or homogeneous) boundary conditions are simply the special cases when the stiffness for each of these (translational and rotational) springs is equal to either zero or infinity.

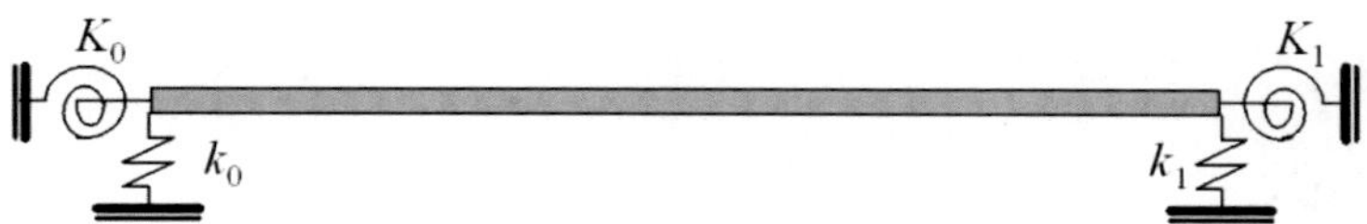

Figure 2. A beam elastically restrained at both ends.

Due to the presence of the elastic supports, the beam is generally allowed to move at both ends. As discussed earlier, while the conventional Fourier method is no longer useful in such a case, the modified one will be directly applicable.

In a form slightly different from Eq. (11), the beam displacement is here expressed as [15]

$$w(x) = \sum_{m=0}^{\infty} A_m \cos\lambda_m x + \xi(x)^T \boldsymbol{\alpha} , \qquad\qquad 0 \leq x \leq L , \qquad\qquad (20)$$

where

$$\boldsymbol{\alpha} = \{\alpha_0, \alpha_1, \beta_0, \beta_1\}^T = \{w'''(0), w'''(L), w'(0), w'(L)\}^T , \qquad\qquad (21)$$

and

$$\xi(x)^T = \begin{cases} -(15x^4 - 60Lx^3 + 60L^2x^2 - 8L^4)/360L \\ (15x^4 - 30L^2x^2 + 7L^4)/360L \\ (6Lx - 2L^2 - 3x^2)/6L \\ (3x^2 - L^2)/6L \end{cases} . \qquad\qquad (22)$$

Substitution of Eq. (20) into the boundary conditions, Eqs. (16-19), results in

$$\hat{k}_0 \left(\sum_{m=0}^{\infty} A_m + \frac{8L^3\alpha_0}{360} + \frac{7L^3\alpha_1}{360} - \frac{\beta_0 L}{3} - \frac{\beta_1 L}{6} \right) = -\alpha_0 , \qquad\qquad (23)$$

$$\hat{k}_1 \left(\sum_{m=0}^{\infty} (-1)^m A_m - \frac{7L^3\alpha_0}{360} - \frac{8L^3\alpha_1}{360} + \frac{\beta_0 L}{6} + \frac{\beta_1 L}{3} \right) = \alpha_1 , \qquad\qquad$$

$$(\hat{k}_0 = k_0/D, \ \hat{k}_1 = k_1/D) \qquad\qquad (24)$$

$$\hat{K}_0 \beta_0 = \left(-\sum_{m=1}^{\infty} \lambda_m^2 A_m - \frac{\alpha_0 L}{3} - \frac{\alpha_1 L}{6} + \frac{\beta_1}{L} - \frac{\beta_0}{L} \right),$$

(25)

and

$$\hat{K}_1 \beta_1 = -\left(\sum_{m=1}^{\infty} (-1)^{m+1} \lambda_m^2 A_m + \frac{\alpha_0 L}{6} + \frac{\alpha_1 L}{3} + \frac{\beta_1}{L} - \frac{\beta_0}{L} \right)$$

$$(\hat{K}_0 = K_0/D, \quad \hat{K}_1 = K_1/D).$$

(26)

Equations (23-26) can be rewritten in matrix form as

$$\boldsymbol{\alpha} = \mathbf{H}^{-1} \sum_{m=0}^{\infty} \mathbf{Q}_m A_m$$

(27)

where

$$\mathbf{H} = \begin{bmatrix} \dfrac{8\hat{k}_0 L^3}{360} + 1 & \dfrac{7\hat{k}_0 L^3}{360} & \dfrac{-\hat{k}_0 L}{3} & \dfrac{-\hat{k}_0 L}{6} \\[2mm] \dfrac{7\hat{k}_1 L^3}{360} & \dfrac{8\hat{k}_1 L^3}{360} + 1 & \dfrac{-\hat{k}_1 L}{6} & \dfrac{-\hat{k}_1 L}{3} \\[2mm] \dfrac{L}{3} & \dfrac{L}{6} & \hat{K}_0 + \dfrac{1}{L} & \dfrac{-1}{L} \\[2mm] \dfrac{L}{6} & \dfrac{L}{3} & \dfrac{-1}{L} & \hat{K}_1 + \dfrac{1}{L} \end{bmatrix}$$

(28)

and

$$\mathbf{Q}_m = \left\{ -\hat{k}_0 \quad (-1)^m \hat{k}_1 \quad -\lambda_m^2 \quad (-1)^m \lambda_m^2 \right\}^T.$$

(29)

When the boundary unknowns, $\boldsymbol{\alpha}$, are determined from Eq. (27), the boundary conditions will all be satisfied automatically regardless of the values for the Fourier coefficients. However, the actual displacement function corresponds to a unique set of Fourier coefficients which have to explicitly satisfy the governing differential equation, Eq. (14). By substituting Eqs. (20) and (27) into (14) and equating the coefficients for the like terms, $\cos \lambda_m x$, on both sides, one can easily obtain

$$\lambda_m^4 A_m - \rho_D \omega^2 \left(A_m + \sum_{m'=0}^{\infty} S_{mm'} A_{m'} \right) = 0 \qquad m = 1, 2, 3, \ldots$$

(30)

and

$$\sum_{m'=0}^{\infty} \mathbf{c} \mathbf{H}^{-1} \mathbf{Q}_{m'} A_{m'} - 2\rho_D \omega^2 A_0 = 0 \tag{31}$$

where

$$\mathbf{c} = \{-2/L \quad 2/L \quad 0 \quad 0\}^T, \quad S_{mm'} = \mathbf{P}_m^T \mathbf{H}^{-1} \mathbf{Q}_{m'}, \tag{32, 33}$$

and

$$\mathbf{P}_m = \frac{2}{L} \left\{ \frac{1}{\lambda_m^{\ 4}} \quad \frac{(-1)^{m+1}}{\lambda_m^{\ 4}} \quad \frac{-1}{\lambda_m^{\ 2}} \quad \frac{(-1)^m}{\lambda_m^{\ 2}} \right\}^T. \tag{34}$$

If the Fourier series is truncated to $m=M$, Eqs. (30) and (31) can be rewritten as

$$\left(\mathbf{K} - \rho_D \omega^2 \mathbf{M}\right) \mathbf{A} = 0 \tag{35}$$

where

$$\mathbf{A} = \{A_0, A_1, ..., A_M\}^T, \tag{36}$$

$$\mathbf{K} = \begin{bmatrix} \mathbf{c}^T \mathbf{H}^{-1} \mathbf{Q}_0 & \mathbf{c}^T \mathbf{H}^{-1} \mathbf{Q}_1 & \cdots & \mathbf{c}^T \mathbf{H}^{-1} \mathbf{Q}_{m'} & \cdots & \mathbf{c}^T \mathbf{H}^{-1} \mathbf{Q}_M \\ 0 & \lambda_1^4 & \cdots & 0 & \cdots & 0 \\ \vdots & \vdots & \cdots & \vdots & \cdots & \vdots \\ 0 & 0 & \cdots & \lambda_m^4 \delta_{mm'} & \cdots & 0 \\ \vdots & \vdots & \cdots & \vdots & \cdots & \vdots \\ 0 & 0 & \cdots & 0 & \cdots & \lambda_M^4 \end{bmatrix}, \tag{37}$$

and

$$\mathbf{M} = \begin{bmatrix} 1 & 0 & \cdots & 0 & \cdots & 0 \\ S_{10} & 1+S_{11} & \cdots & S_{1m'} & \cdots & S_{1M} \\ \vdots & \vdots & \cdots & \vdots & \cdots & \vdots \\ S_{m0} & S_{m1} & \cdots & \delta_{mm'} + S_{mm'} & \cdots & S_{mM} \\ \vdots & \vdots & \cdots & \vdots & \cdots & \vdots \\ S_{M0} & S_{M1} & \cdots & S_{Mm'} & \cdots & 1+S_{MM} \end{bmatrix}. \tag{38}$$

It is interesting to note from Eq. (30) that the current formulation has lead to a different interpretation of the physical impact of the boundary springs: instead of affecting the stiffness matrix, they actually change the effective mass of the beam.

The natural frequencies and eigenvectors can now be easily obtained by solving a standard matrix eigenproblem. The eigenvectors are actually the Fourier coefficients from which the corresponding mode shapes can be readily determined from

$$w(x) = \sum_{m=0}^{\infty} A_m (\cos\lambda_m x + \boldsymbol{\xi}(x)^T \mathbf{H}^{-1}\mathbf{Q}_m) \qquad . \tag{39}$$

This equation can be alternatively expressed as

$$w(x) = \sum_{m=0}^{\infty} A_m \phi_m(x) \tag{40}$$

where

$$\phi_m(x) = \cos\lambda_m x + \boldsymbol{\xi}(x)^T \mathbf{H}^{-1}\mathbf{Q}_m \quad . \tag{41}$$

Equation (41) essentially defines a new set of basis functions, $\{\phi_m(x), m = 0, 1, 2, 3, ...\}$, which satisfy all the specified boundary conditions, Eqs. (16-19). This interpretation allows an alternative way of solving the governing differential equation based on the Galerkin procedure [16]. The resulting equations can be accordingly written as

$$\sum_{m'=1}^{\infty} (\delta_{mm'} + S_{m'm})\lambda_{m'}^4 A_{m'} - \rho_D\omega^2 \left((S_{m0} + Z_{m0})A_0 + \sum_{m'=1}^{\infty} (\delta_{mm'} + S_{mm'} + S_{m'm} + Z_{mm'})A_{m'} \right) = 0 \; ,$$
$$m = 1, 2, 3, ... \tag{42}$$

and

$$\mathbf{cH}^{-1}\mathbf{Q}_0 A_0 + \sum_{m'=1}^{\infty} (\mathbf{cH}^{-1}\mathbf{Q}_{m'} + \lambda_{m'}^4 S_{m'0})A_{m'} - \rho_D\omega^2 \left((2 + Z_{00})A_0 + \sum_{m'=1}^{\infty} (S_{m'0} + Z_{0m'})A_{m'} \right) = 0 \tag{43}$$

where

$$Z_{mm'} = \mathbf{Q}_m^T \mathbf{H}^{-T}\boldsymbol{\Xi}\mathbf{H}^{-1}\mathbf{Q}_{m'} \tag{44}$$

and

$$\boldsymbol{\Xi} = 2/L \int_0^L \boldsymbol{\xi}(x)^T \boldsymbol{\xi}(x)\,dx$$

$$= \begin{bmatrix} \dfrac{2L^6}{4725} & & & \\[2ex] \dfrac{127L^6}{302400} & \dfrac{2L^6}{4725} & & sym. \\[2ex] -\dfrac{4L^4}{945} & -\dfrac{31L^4}{7560} & \dfrac{2L^2}{45} & \\[2ex] -\dfrac{31L^4}{7560} & -\dfrac{4L^4}{945} & \dfrac{7L^2}{180} & \dfrac{2L^2}{45} \end{bmatrix} . \tag{45}$$

In addition, it is not difficult to verify that

$$S_{m'm}\lambda_{m'}^4 = S_{mm'}\lambda_m^4 , \qquad\qquad Z_{m'm} = Z_{mm'} \quad , \tag{46, 47}$$

and

$$\mathbf{cH}^{-1}\mathbf{Q}_{m'} + \lambda_{m'}^4 S_{m'0} \equiv 0 . \tag{48}$$

Making use of Eq. (48), (43) reduces to

$$\mathbf{cH}^{-1}\mathbf{Q}_0 A_0 - \rho_D\omega^2\left((2 + Z_{00})A_0 + \sum_{m'=1}^{\infty}\left(S_{m'0} + Z_{m'0}\right)A_{m'} \right) = 0 \tag{49}$$

Finally, Eqs. (42) and (49) can be combined as

$$\left(\mathbf{K} - \rho_D\omega^2\mathbf{M}\right)\mathbf{A} = 0 \tag{50}$$

where

$$K_{mm'} = (1-\delta_{0m})(1-\delta_{0m'})(\delta_{mm'} + S_{m'm})\lambda_{m'}^4 + \delta_{0m'}\delta_{m0}(\mathbf{cH}^{-1}\mathbf{Q}_0) , \tag{51}$$

and

$$M_{mm'} = (1-\delta_{0m})(1-\delta_{0m'})(\delta_{mm'} + S_{mm'} + S_{m'm} + Z_{mm'}) + \delta_{0m'}\delta_{m0}(2 + Z_{00})$$

$$+ \delta_{0m'}(S_{m0} + Z_{m0}) + \delta_{m0}(S_{m'0} + Z_{m'0}) ,$$

$$\text{for } m, m' = 0, 1, 2, 3, \ldots \tag{52}$$

In comparison with Eqs. (37) and (38) in the Fourier series method, the Galerkin scheme has led to some additional terms in the stiffness and mass matrices as evident from Eqs. (51) and (52). Although it is not immediately clear if the additional calculations have any practical benefits, the stiffness and mass matrices now become symmetric, which is a desired feature numerically.

The displacement solution can also be expanded into a sine series [17]

$$w(x) = \sum_{m=0}^{\infty} A_m \sin \lambda_m x + \xi(x)^T \alpha, \qquad\qquad 0 \le x \le L. \tag{53}$$

However, the following modifications have to be made accordingly:

$$\alpha = \{\alpha_0, \alpha_1, \beta_0, \beta_1\}^T = \{w''(0), w''(L), w(0), w(L)\}^T, \tag{54}$$

$$\xi(x) = \left\{ \begin{array}{c} -(2L^2 x - 3Lx^2 + x^3)/6L \\ (x^3 - L^2 x)/6L \\ (L - x)/L \\ x/L \end{array} \right\}, \tag{55}$$

$$\mathbf{H} = \begin{bmatrix} \dfrac{\hat{K}_0 L}{3} + 1 & \dfrac{\hat{K}_0 L}{6} & \dfrac{\hat{K}_0}{L} & -\dfrac{\hat{K}_0}{L} \\[2mm] \dfrac{\hat{K}_1 L}{6} & \dfrac{\hat{K}_1 L}{3} + 1 & -\dfrac{\hat{K}_1}{L} & \dfrac{\hat{K}_1}{L} \\[2mm] -\dfrac{1}{L} & \dfrac{1}{L} & \hat{k}_0 & 0 \\[2mm] \dfrac{1}{L} & -\dfrac{1}{L} & 0 & \hat{k}_1 \end{bmatrix}, \tag{56}$$

and

$$\mathbf{Q}_m = \left\{ \hat{K}_0 \lambda_m \quad (-1)^{m+1} \hat{K}_1 \lambda_m \quad \lambda_m^3 \quad (-1)^{m+1} \lambda_m^3 \right\}^T. \tag{57}$$

The final system equation can be written as

$$\lambda_m^4 A_m - \rho_D \omega^2 \left(A_m + \sum_{m'=1}^{\infty} S_{mm'} A_{m'} \right) = 0 \qquad m = 1, 2, 3, \ldots \tag{58}$$

where $S_{mm'}$ is still defined by Eq. (33), but

$$\mathbf{P}_m = \frac{2}{L} \left\{ \frac{-1}{\lambda_m^3} \quad \frac{(-1)^m}{\lambda_m^3} \quad \frac{1}{\lambda_m} \quad \frac{(-1)^{m+1}}{\lambda_m} \right\}^T. \tag{59}$$

Thus far, the discussions have been primarily focused on how to ensure the series expansions to correctly converge to the actual boundary values specified at both ends in beam problems. Mathematically, it is extremely important to fully understand the convergence characteristic of any series solution. For example, it has been shown that the displacement function can be expanded into both cosine and sine series. Then a natural question is whether there is any difference between them in terms of solution accuracy and convergence. To answer this question, let us quote an importance convergence theorem [14]:

Theorem 3. *Let $f(x)$ be a continuous function of period 2L, which has m derivatives, where m-1 derivatives are continuous and the m-th derivative is absolutely integrable (the m'th derivative may not exist at certain points). Then, the Fourier series of all m derivatives can be obtained by term-by-term differentiation of the Fourier series of $f(x)$, where all the series, except possibly the last, converge to the corresponding derivatives. Moreover, the Fourier coefficients of the function $f(x)$ satisfy the relations*

$$\lim_{n \to \infty} a_n \lambda_n^m = \lim_{n \to \infty} b_n \lambda_n^m = 0. \tag{60}$$

Because the Fourier series now represents a residual displacement function which has at least three continuous derivatives, one can anticipate, according to Eq. (60), the resulting series expansions, Eqs. (20) and (53), will both converge at a remarkable speed of

$$\lim_{n \to \infty} A_n \lambda_n^4 = 0. \tag{61}$$

Actually, for a generally supported beam, the convergence of the Fourier series solutions can be directly estimated from [17]

$$A_m = \frac{\rho_D \omega^2 P_m}{\lambda_m^4 - \rho_D \omega^2} = \frac{2 \rho_D \omega^2 / \lambda_m^2}{(\lambda_m^4 - \rho_D \omega^2)L} \left(-\frac{\alpha_1 (-1)^m - \alpha_0}{\lambda_m^2} + \beta_1 (-1)^m - \beta_0 \right) \tag{62}$$

for cosine series expansion;

$$A_m = \frac{\rho_D \omega^2 P_m}{\lambda_m^4 - \rho_D \omega^2} = \frac{2 \rho_D \omega^2 / \lambda_m}{(\lambda_m^4 - \rho_D \omega^2)L} \left(\frac{\alpha_1 (-1)^m - \alpha_0}{\lambda_m^2} + \beta_1 (-1)^{m+1} + \beta_0 \right) \tag{63}$$

for sine series expansion.

It should be noted that although the same symbols are used in Eqs. (62) and (63), the boundary constants, α_i and β_i ($i=0,1$), actually have different meanings (refer to Eqs. (21) and (54)). Suppose that the boundary constants are somehow known *a priori*, then Eqs. (62) and (63) can be respectively used to assess the convergence rates of the cosine and sine series expansions. For instance, if a beam is simply supported at each end with only rotational restraint, the constants β_0 and β_1 (representing the displacements at $x=0$ and L) in the sine series expansion are then both equal to zero. However, this is not generally true for the cosine series representation because there these constants denote the boundary values of the first derivatives. According to Eqs. (62) and (63), one can expect that the sine series converges at a rate of $A_m \sim O(\lambda_m^{-7})$ and the cosine series $A_m \sim O(\lambda_m^{-6})$. However, the cosine series will converge much faster if the first derivative vanishes at $x=0$ and L as in the cases when a beam is guided at each end with only translational restraint. In such a situation, we have that $A_m \sim O(\lambda_m^{-5})$ for the sine series and $A_m \sim O(\lambda_m^{-8})$ for the cosine series. For beams with general elastic restraints, the constants β_0 and β_1 will not be both equal to zero typically. Therefore, the sine and cosine series will respectively converge according to $A_m \sim O(\lambda_m^{-5})$ and $A_m \sim O(\lambda_m^{-6})$. In other words, the cosine series will typically outperform its sine counterpart for generally supported beams in terms of the rate of convergence.

EXAMPLE 1. First consider a beam clamped at $x=0$, and simply supported with an elastic rotational constraint at $x=L$. The condition that a displacement or rotation is zero at an end is equivalent to setting the stiffness of the corresponding spring to infinity (say, 10^{10} in actual numerical calculations). Table 1 shows the first eight frequency parameters, $\mu_i = L/\pi(\omega_i\sqrt{\rho A/D})^{1/2}$, for the various stiffnesses at the simply supported end. For the two extreme stiffnesses, $\hat{K}_1 L = 0$ and 10^{10}, this problem reduces to the classical clamped-pinned and clamped-clamped cases for which the first four frequency parameters are well known as [18]: μ_i=1.24988, 2.25, 3.25, 4.25 and μ_i=1.50562, 2.49975, 3.50001, 4.5, respectively.

Table 1. Frequency parameters, $\mu_i = L/\pi(\omega_i\sqrt{\rho A/D})^{1/2}$, for various stiffnesses of the rotational springs

Mode	$\mu_i = L/\pi(\omega_i\sqrt{\rho A/D})^{1/2}$				
	$\hat{K}_1 L = 0$	$\hat{K}_1 L = 1$	$\hat{K}_1 L = 10$	$\hat{K}_1 L = 100$	$\hat{K}_1 L = 10^{10}$
1	1.24988	1.28656	1.4102	1.49137	1.50562
2	2.25005	2.27081	2.37138	2.47681 3.46884	2.49975
3	3.25014	3.26491	3.34927	4.46108	3.50001
4	4.25032	4.26175	4.3337		4.5

The frequencies in Table 1 are obtained by truncating the Fourier cosine series, Eq. (20), to M=20. To examine the convergence of the series solution, Table 2 compares the first ten frequency parameters (for the clamped-clamped beam) calculated by using different numbers of terms in the Fourier series expansion. In Figure 3, the mode shapes are plotted for two of the lowest modes for the clamped-clamped beam. The classical solution for this case is well known as [18].

Table 2. Frequency parameters, $\mu_i = L/\pi(\omega_i\sqrt{\rho A/D})^{1/2}$, for various numbers of terms in Fourier cosine series

Mode	$\mu_i = L/\pi(\omega_i\sqrt{\rho A/D})^{1/2}$			
	$M=5$	$M=10$	$M=15$	$M=20$
1	1.50563	1.50562	1.50562	1.50562
2	2.49985	2.49976	2.49975	2.49975
3	3.50392	3.50003	3.50001	3.50001
4	4.5073	4.5002	4.50001	4.5
5		5.50044	5.50005	5.5
6		6.50289	6.5001	6.50002
7		7.50421	7.50045	7.50004
8		8.52423	8.5007	8.50014
9		9.52852	9.50251	9.50022
10			10.5033	10.5007

$$\phi_i = \cosh\frac{\pi\mu_i x}{L} - \cos\frac{\pi\mu_i x}{L} - \sigma_i\left(\sinh\frac{\pi\mu_i x}{L} - \sin\frac{\pi\mu_i x}{L}\right) \tag{64}$$

where

$$\sigma_i = \frac{\cosh\pi\mu_i - \cos\pi\mu_i}{\sinh\pi\mu_i - \sin\pi\mu_i}. \tag{65}$$

In each of the plots, M denotes the number of cosine terms actually used in the Fourier series, and the curves for larger M's are not presented simply because they then become virtually identical to the classical solution. All these results have indicated that the mode shapes can also be accurately obtained by taking only a few terms in the Fourier series.

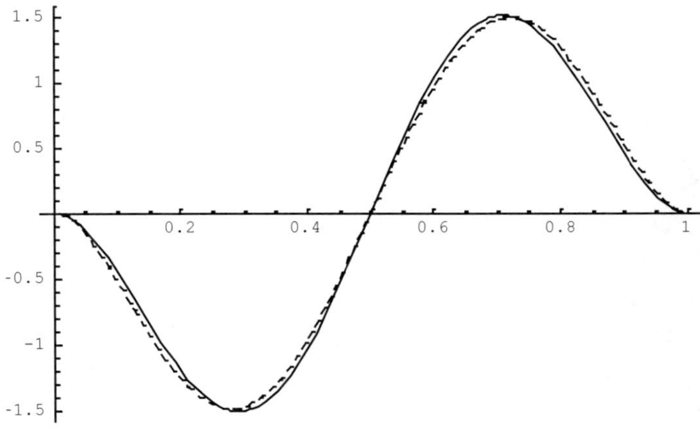

Figure 3 (Continued)

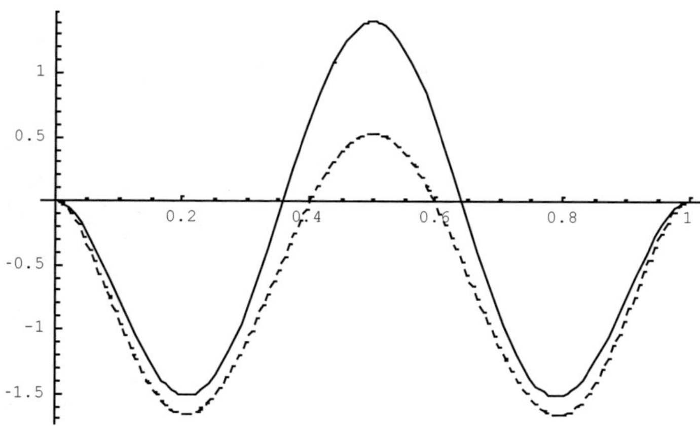

Figure 3. The mode shapes for: (a) the second mode, , Eq. (64), , Eq. (20), M=2; (b) the third mode, , Eq. (64), , Eq. (20), M=3.

Table 3. Fundamental natural frequency, $\mu_1 = (L^2\omega_1\sqrt{\rho A/D})^{1/2}$, of a beam with various combinations of the rotational and translational springs

$\hat{K}_0 L$		$\hat{k}_1 L^3$	
	0.01	1	100
0.01	0.4948	1.3134	2.9901
1	1.2520	1.5358	3.1085
100	1.8583	1.9940	3.6134

EXAMPLE 2. Let's now consider a problem that involves both rotational and translational elastic restraints. Assume the beam is pinned at the left end with a rotational spring and a translational spring is attached to the other end. Table 3 lists the (dimensionless) fundamental frequencies of the beam for the various combinations of the spring constants. The results are almost exactly the same as those given by Maurizi *et al.* [6].

VIBRATIONS OF TWO ARBITRARILY COUPLED BEAMS

Many analytical methods are available in the literature for modeling two or more beams coupled together in various ways. Of them are the popular Dynamic Stiffness Methods (DSM) [19-21], Spectral Element Method (SEM) [2-4], and receptance theory [22-26]. In the DSM, the displacements at the ends of a beam are determined in terms of the complex amplitudes for each wave component from which one can derive the relationship between the general force and displacement vectors. In SEM each uniform beam can be considered as a super-element on which the flexural, longitudinal and torsional waves will be expressed in terms of exact beam solutions. In the receptance methods, the relationship between the coupling forces and the displacements at the ends of a beam is derived by making use of the Green's functions for the uncoupled beam under boundary conditions *compatible* with the entire system. The Green's function can be expressed either in a closed-form or as the expansion of the modes for the beam with appropriate boundary conditions.

Although these techniques have been widely used for the dynamic analysis of coupled beams or beam systems, there are various technical and practical issues and concerns related to each of them. For instance, the applications of SEM and DSM are primarily limited to beam frameworks because there is no exact form of general solutions for other types of structural components such as plates. In addition, the beams have to present certain degree of uniformity regarding the material and geometrical properties; for example, a beam cannot have a varying mass density or cross-sectional area. In dealing with a beam structure, the system solution is often expressed in terms of the modal properties for each individual beam by assuming the coupling end(s) is free. The modal properties such obtained may also be alternatively used to determine the transfer functions, Green functions or receptance functions between the responses and the reaction forces (including moments) at the junctions in the actual system environment. While this solution procedure may be reasonably good for calculating the modal properties or the vibrational response of the composite structure to an external load, the basic fact still cannot be changed; that is, the modal properties for each beam are actually determined by freeing the "coupling" end(s). As a consequence, the displacement functions constructed using these component modes will not be able to accurately represent the reaction forces/moments (which depend upon the second and third derivatives of the displacement functions) at the coupling ends. This problem is expected to become more remarked in a power flow analysis because the reaction forces and moments will have to be calculated explicitly at the junctions. Although for an elastic joint the coupling forces and moments can be respectively calculated from the relative translational and rotational displacements, the calculations will understandably break down when the coupling stiffnesses become very high. The use of free-free beam functions was also found difficult for the intermediate coupling strength [27]. Other well-recognized problems or concerns include: the numerical break-down of the standard beam functions approximately after the first dozen modes, the slow convergence of the modal expansions when the boundary/coupling conditions cannot be assumed correctly, and so on. In view of the aforementioned issues and concerns, the development of a robust and sophisticated method which is capable of simplifying solution algorithms, reducing model input data, and universally dealing with various coupling and boundary conditions will be of great interest to both researchers and application engineers. It will be evident from the following discusses that all these problems and concerns can be satisfactorily resolved by the modified Fourier solution method.

Figure 4 shows two beams which are connected together at an arbitrary angle θ. A set of three linear and rotational springs is used here to represent a general coupling between these two beams. The finite stiffness values for the springs will describe a flexible joint between the beams. The rigid connection is essentially obtained by setting the spring stiffnesses substantially larger than the bending rigidities of the involved beams. Similarly, the beams are assumed to be elastically restrained at the ends. Thus, any classical homogeneous boundary conditions can be considered as a special case when the stiffnesses of these elastic restraints become extremely large or small.

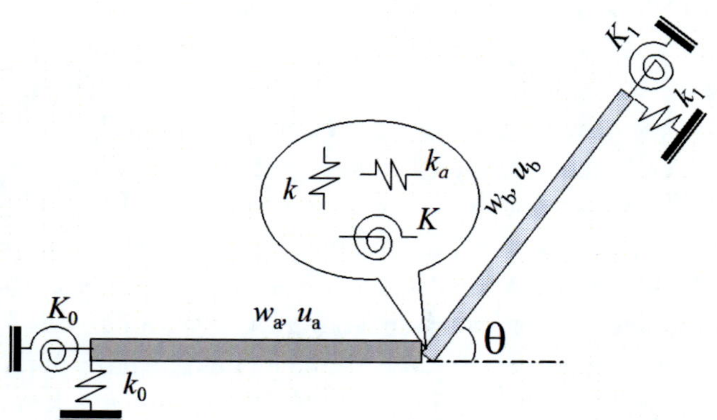

Figure 4. Two beams connected at an arbitrary angle.

The differential equation for the transverse vibration of a beam is

$$D_i\, d^4 w_i(x)/d x^4 - \rho_i S_i \omega^2 w_i(x) = f_{wi}(x), \qquad (i = A, B) \tag{66}$$

where w_i, D_i, ρ_i and S_i are respectively the flexural displacement, the bending rigidity, the mass density and the cross-sectional area of beam i; f_{wi} is the distributed load acting on beam i.

It is evident from Figure 4 that for an angled connection the transverse vibration of beam A will be directly coupled with both the transverse and axial displacements for beam B, and *vice versa*. Thus, the longitudinal vibration of each beam must also be taken into account. The differential equation for the longitudinal displacement of a beam is

$$E_i S_i\, d^2 u_i(x)/d x^2 + \rho_i S_i \omega^2 u_i(x) = f_{ui}(x) \qquad (i = A, B) \tag{67}$$

where u_i and E_i are respectively the longitudinal displacement and Young's modulus for beam i and f_{ui} is the axial distributed load.

The boundary and coupling conditions are as follows:

$$k_0 w_A(0) = -D_A w_A'''(0), \tag{68}$$

$$K_0 w_A'(0) = D_A w_A''(0), \tag{69}$$

$$k_1 w_B(L_B) = D_B w_B'''(L_B), \tag{70}$$

$$K_1 w_B'(L_B) = -D_B w_B''(L_B), \tag{71}$$

$$D_A w_A'''(L_A) = k[w_A(L_A) - w_B(0)] , \tag{72}$$

$$D_A w_A''(L_A) = -K[w_A'(L_A) - w_B'(0)\cos\theta + u_B(0)\sin\theta] , \tag{73}$$

$$D_B w_B'''(0) = -k[w_B(0) - w_A(L_A)] ,$$
(74)

$$D_B w_B''(0) = K \cos^2 \theta [w_B'(0) - w_A'(L_A)\cos\theta) + K_a \sin^2 \theta (w_B'(0) - u_A(L_A)\sin\theta] ,$$
(75)

$$E_A S_A u_A'(0) = k_{a0} u_A(0)$$
(76)

$$E_B S_B u_B'(L_B) = -k_{a1} u_B(L_B)$$
(77)

$$E_A S_A u_A'(L_A) = -k_a [u_A(L_A) - u_B(0)\cos\theta + w_B(0)\sin\theta]$$
(78)

$$E_B S_B u_B'(0) = k_a \cos^2 \theta [u_B(0) - u_A(L_A)\cos\theta] + k \sin^2 \theta [u_B(0) - w_A(L_A)\sin\theta] \quad (79)$$

where k, K and k_a are the stiffnesses for the coupling springs (refer to Figure 4) and those with subscripts 0 and 1 denote the stiffnesses for the springs at the left and right end of the system, respectively.

In light of Eq. (20), the transverse and axial displacements on beam i will be, respectively, sought as

$$w_i(x) = \sum_{m=0}^{\infty} C_{wi}^m \cos\lambda_{im}x + \phi_{wi}(x) ,$$
(80)

and

$$u_i(x) = \sum_{m=0}^{\infty} C_{ui}^m \cos\lambda_{im}x + \phi_{ui}(x) , \quad 0 \leq x \leq L_i$$
(81)

where L_i is the length of beam i.

The auxiliary functions $\phi_{wi}(x)$ for the transverse displacement are required, *regardless of the actual boundary/coupling conditions*, to satisfy Eq. (21) where the values of derivatives should now be understood as those at the both ends of beam i. Because the first derivative of the axial displacement is required to be continuous, the auxiliary functions $\phi_{ui}(x)$ only needs to satisfy the following conditions:

$$\phi_{ui}'(0) = u_i'(0) = \beta_{wi0} ,$$
(82)

and

$$\phi_{ui}'(L_i) = u_i'(L_i) = \beta_{wi1} .$$
(83)

Such an auxiliary function $\phi_{wi}(x)$ can be easily expressed, for example, as

$$\phi_{ui} = \xi_{ui}(x)^T \alpha_{ui} \tag{84}$$

where

$$\alpha_{ui} = \{\beta_{ui0}, \beta_{ui1}\}^T, \tag{85}$$

and

$$\xi_{ui}(x)^T = \left\{ \begin{array}{c} (6L_i x - 2L_i^2 - 3x^2)/6L_i \\ (3x^2 - L_i^2)/6L_i \end{array} \right\}. \tag{86}$$

It should be noted that although the same symbol x is used in Eqs. (80) and (81), it actually represents two different (local) coordinate systems whose origins are located at the left ends of the beams, respectively.

Substitution of Eqs. (80) and (81) into the boundary and coupling conditions, Eqs. (68-79), will lead to

$$k_0 \left(\sum_{m=0}^{\infty} C_{wA}^m + \frac{8L_A^3 \alpha_{wA0}}{360} + \frac{7L_A^3 \alpha_{wA1}}{360} - \frac{\beta_{wA0}L_A}{3} - \frac{\beta_{wA1}L_A}{6} \right) = -D_A \alpha_{wA0}, \tag{87}$$

$$K_0 \beta_{wA0} = D_A \left(-\sum_{m=1}^{\infty} \lambda_{Am}^2 C_{wA}^m - \frac{\alpha_{wA0}L_A}{3} - \frac{\alpha_{wA1}L_A}{6} + \frac{\beta_{wA1}}{L_A} - \frac{\beta_{wA0}}{L_A} \right), \tag{88}$$

$$k_1 \left(\sum_{m=0}^{\infty} (-1)^m C_{wB}^m - \frac{7L_B^3 \alpha_{wB0}}{360} - \frac{8L_B^3 \alpha_{wB1}}{360} + \frac{\beta_{wB0}L_B}{6} + \frac{\beta_{wB1}L_B}{3} \right) = D_B \alpha_{wB1}, \tag{89}$$

$$K_1 \beta_{wB1} = -D_B \left(\sum_{m=1}^{\infty} (-1)^{m+1} \lambda_{Bm}^2 C_{wB}^m + \frac{\alpha_{wB0}L_B}{6} + \frac{\alpha_{wB1}L_B}{3} + \frac{\beta_{wB1}}{L_B} - \frac{\beta_{wB0}}{L_B} \right), \tag{90}$$

$$k \left(\sum_{m=0}^{\infty} (-1)^m C_{wA}^m - \frac{7L_A^3 \alpha_{wA0}}{360} - \frac{8L_A^3 \alpha_{wA1}}{360} + \frac{\beta_{wA0}L_A}{6} + \frac{\beta_{wA1}L_A}{3} \right) =$$

$$D_A \alpha_{A1} + k\cos\theta \left(\sum_{m=0}^{\infty} C_{wB}^m + \frac{8L_B^3 \alpha_{B0}}{360} + \frac{7L_B^3 \alpha_{B1}}{360} - \frac{\beta_{B0}L_B}{3} - \frac{\beta_{B1}L_B}{6} \right)$$

$$+ k\sin\theta \left(\sum_{m=0}^{\infty} C_{uB}^m - \frac{\beta_{B0}L_B}{3} - \frac{\beta_{B1}L_B}{6} \right), \tag{91}$$

$$K\beta_{wB0} = K\beta_{wA1} + D_B \left(-\sum_{m=1}^{\infty} \lambda_{Bm}^2 C_{wB}^m - \frac{\alpha_{wB0}L_B}{3} - \frac{\alpha_{wB1}L_B}{6} + \frac{\beta_{wB1}}{L_B} - \frac{\beta_{wB0}}{L_B} \right), \tag{92}$$

$$D_B \alpha_{wB0} + (k \cos^2 \theta + k_a \sin^2 \theta) \left(\sum_{m=0}^{\infty} C_{wB}^m + \frac{8 L_B^3 \alpha_{wB0}}{360} + \frac{7 L_B^3 \alpha_{wB1}}{360} - \frac{\beta_{wB0} L_B}{3} - \frac{\beta_{wB1} L_B}{6} \right) =$$

$$k \, \cos^3 \theta \left(\sum_{m=0}^{\infty} (-1)^m C_{wA}^m - \frac{7 L_A^3 \alpha_{wA0}}{360} - \frac{8 L_A^3 \alpha_{wA1}}{360} + \frac{\beta_{wA0} L_A}{6} + \frac{\beta_{wA1} L_A}{3} \right)$$

$$- k_a \sin^3 \theta \left(\sum_{m=0}^{\infty} (-1)^m C_{uA}^m + \frac{\beta_{uA0} L_A}{6} + \frac{\beta_{uA1} L_A}{3} \right), \tag{93}$$

$$K \beta_{wA1} = K \beta_{wB0} - D_A \left(\sum_{m=1}^{\infty} (-1)^{m+1} \lambda_{Am}^2 C_{wA}^m + \frac{\alpha_{wA0} L_A}{6} + \frac{\alpha_{wA1} L_A}{3} + \frac{\beta_{wA1}}{L_A} - \frac{\beta_{wA0}}{L_A} \right), \tag{94}$$

$$k_{a0} \left(\sum_{m=0}^{\infty} C_{uA}^m - \frac{\beta_{uA0} L_A}{3} - \frac{\beta_{u1} L_A}{6} \right) = E_A S_A \beta_{uA0}, \tag{95}$$

$$k_{a1} \left(\sum_{m=0}^{\infty} (-1)^m C_{uB}^m + \frac{\beta_{uB0} L_B}{6} + \frac{\beta_{uB1} L_B}{3} \right) = -E_B S_B \beta_{uB1}, \tag{96}$$

$$k_a \left(\sum_{m=0}^{\infty} (-1)^m C_{uA}^m + \frac{\beta_{uA0} L_A}{6} + \frac{\beta_{uA1} L_A}{3} \right) + E_A S_A \beta_{uA1} =$$

$$- k_a \sin \theta \left(\sum_{m=0}^{\infty} C_{wB}^m + \frac{8 L_B^3 \alpha_{wB0}}{360} + \frac{7 L_B^3 \alpha_{wB1}}{360} - \frac{\beta_{wB0} L_B}{3} - \frac{\beta_{wB1} L_B}{6} \right)$$

$$+ k_a \cos \theta \left(\sum_{m=0}^{\infty} C_{uB}^m - \frac{\beta_{uB0} L_B}{3} - \frac{\beta_{uB1} L_B}{6} \right), \tag{97}$$

and

$$E_B S_B \beta_{uB0} - (k_a \cos^2 \theta + k \sin^2 \theta) \left(\sum_{m=0}^{\infty} C_{uB}^m - \frac{\beta_{uB0} L_B}{3} - \frac{\beta_{uB1} L_B}{6} \right) =$$

$$- k \, \sin^3 \theta \left(\sum_{m=0}^{\infty} (-1)^m C_{wA}^m - \frac{7 L_A^3 \alpha_{wA0}}{360} - \frac{8 L_A^3 \alpha_{wA1}}{360} + \frac{\beta_{wA0} L_A}{6} + \frac{\beta_{wA1} L_A}{3} \right)$$

$$- k_a \cos^3 \theta \left(\sum_{m=0}^{\infty} (-1)^m C_{uA}^m + \frac{\beta_{uA0} L_A}{6} + \frac{\beta_{uA1} L_A}{3} \right). \tag{98}$$

From equations (87-98) the unknown boundary parameters can be found as

$$\overline{\alpha} = \left\{ \begin{array}{c} \alpha_{wA} \\ \alpha_{wB} \\ \alpha_{uA} \\ \alpha_{uB} \end{array} \right\} = \sum_{m=0}^{\infty} \left\{ \begin{array}{c} \widetilde{\mathbf{H}}_1 \mathbf{Q}_1^m C_{wA}^m + \widetilde{\mathbf{H}}_1 \mathbf{Q}_2^m C_{wA}^m + \widetilde{\mathbf{H}}_1 \mathbf{Q}_3^m C_{uA}^m + \widetilde{\mathbf{H}}_1 \mathbf{Q}_4^m C_{uB}^m \\ \widetilde{\mathbf{H}}_2 \mathbf{Q}_1^m C_{wA}^m + \widetilde{\mathbf{H}}_2 \mathbf{Q}_2^m C_{wA}^m + \widetilde{\mathbf{H}}_2 \mathbf{Q}_3^m C_{uA}^m + \widetilde{\mathbf{H}}_2 \mathbf{Q}_4^m C_{uB}^m \\ \widetilde{\mathbf{H}}_3 \mathbf{Q}_1^m C_{wA}^m + \widetilde{\mathbf{H}}_3 \mathbf{Q}_2^m C_{wA}^m + \widetilde{\mathbf{H}}_3 \mathbf{Q}_3^m C_{uA}^m + \widetilde{\mathbf{H}}_3 \mathbf{Q}_4^m C_{uB}^m \\ \widetilde{\mathbf{H}}_4 \mathbf{Q}_1^m C_{wA}^m + \widetilde{\mathbf{H}}_4 \mathbf{Q}_2^m C_{wA}^m + \widetilde{\mathbf{H}}_4 \mathbf{Q}_3^m C_{uA}^m + \widetilde{\mathbf{H}}_4 \mathbf{Q}_4^m C_{uB}^m \end{array} \right\} \tag{99}$$

where

$$\widetilde{\mathbf{H}} = \left\{ \begin{array}{c} \widetilde{\mathbf{H}}_1 \\ \widetilde{\mathbf{H}}_2 \\ \widetilde{\mathbf{H}}_3 \\ \widetilde{\mathbf{H}}_4 \end{array} \right\} = \mathbf{H}^{-1} \tag{100}$$

with

$$\mathbf{H} = \begin{bmatrix} \mathbf{H}_{11} & \mathbf{H}_{12} & \mathbf{0} & \mathbf{H}_{14} \\ \mathbf{H}_{21} & \mathbf{H}_{22} & \mathbf{H}_{23} & \mathbf{0} \\ \mathbf{0} & \mathbf{H}_{32} & \mathbf{H}_{33} & \mathbf{H}_{34} \\ \mathbf{H}_{31} & \mathbf{0} & \mathbf{H}_{43} & \mathbf{H}_{44} \end{bmatrix}. \tag{101}$$

The detailed expressions for H_{ij} and Q_i ($i, j = 1, 2, 3, 4$) are given in Appendix A.
Making use of Eq. (99), the flexural and axial displacements can be finally expressed as

$$w_A(x) = \sum_{m=0}^{\infty} C_{wA}^m (\cos\lambda_{Am} x + \boldsymbol{\xi}_{wA}(x)^T \widetilde{\mathbf{H}}_1 \mathbf{Q}_1^m) + \sum_{m=0}^{\infty} C_{wB}^m \boldsymbol{\xi}_{wA}(x)^T \widetilde{\mathbf{H}}_1 \mathbf{Q}_2^m + \sum_{m=0}^{\infty} C_{uB}^m \boldsymbol{\xi}_{wA}(x)^T \widetilde{\mathbf{H}}_1 \mathbf{Q}_4^m, \tag{102}$$

$$w_B(x) = \sum_{m=0}^{\infty} C_{wB}^m (\cos\lambda_{Bm} x + \boldsymbol{\xi}_{wB}(x)^T \widetilde{\mathbf{H}}_2 \mathbf{Q}_2^m) + \sum_{m=0}^{\infty} C_{wA}^m \boldsymbol{\xi}_{wB}(x)^T \widetilde{\mathbf{H}}_2 \mathbf{Q}_1^m + \sum_{m=0}^{\infty} C_{uA}^m \boldsymbol{\xi}_{wB}(x)^T \widetilde{\mathbf{H}}_2 \mathbf{Q}_3^m \tag{103}$$

$$u_A(x) = \sum_{m=0}^{\infty} C_{uA}^m (\cos\lambda_{Am} x + \boldsymbol{\xi}_{uA}(x)^T \widetilde{\mathbf{H}}_3 \mathbf{Q}_3^m) + \sum_{m=0}^{\infty} C_{wB}^m \boldsymbol{\xi}_{uA}(x)^T \widetilde{\mathbf{H}}_3 \mathbf{Q}_2^m + \sum_{m=0}^{\infty} C_{uB}^m \boldsymbol{\xi}_{uA}(x)^T \widetilde{\mathbf{H}}_3 \mathbf{Q}_4^m \tag{104}$$

and

$$u_B(x) = \sum_{m=0}^{\infty} C_{uB}^m (\cos\lambda_{Bm} x + \boldsymbol{\xi}_{uB}(x)^T \widetilde{\mathbf{H}}_4 \mathbf{Q}_4^m) + \sum_{m=0}^{\infty} C_{wA}^m \boldsymbol{\xi}_{uB}(x)^T \widetilde{\mathbf{H}}_4 \mathbf{Q}_1^m + \sum_{m=0}^{\infty} C_{uA}^m \boldsymbol{\xi}_{uB}(x)^T \widetilde{\mathbf{H}}_4 \mathbf{Q}_3^m. \tag{105}$$

Substituting Eqs. (102-105) into Eqs. (66) and (67) and following the standard Galerkin's procedure, one will be able to obtain

$$\begin{bmatrix} \mathbf{K}_{11} & \mathbf{K}_{12} & \mathbf{0} & \mathbf{K}_{14} \\ \mathbf{K}_{21} & \mathbf{K}_{22} & \mathbf{K}_{23} & \mathbf{0} \\ \mathbf{0} & \mathbf{K}_{32} & \mathbf{K}_{33} & \mathbf{K}_{34} \\ \mathbf{K}_{41} & \mathbf{0} & \mathbf{K}_{43} & \mathbf{K}_{44} \end{bmatrix} \begin{Bmatrix} \mathbf{C}_{wA} \\ \mathbf{C}_{wB} \\ \mathbf{C}_{uA} \\ \mathbf{C}_{uB} \end{Bmatrix} - \omega^2 \begin{bmatrix} \mathbf{M}_{11} & \mathbf{M}_{12} & \mathbf{0} & \mathbf{M}_{14} \\ \mathbf{M}_{21} & \mathbf{M}_{22} & \mathbf{M}_{23} & \mathbf{0} \\ \mathbf{0} & \mathbf{M}_{32} & \mathbf{M}_{33} & \mathbf{M}_{34} \\ \mathbf{M}_{41} & \mathbf{0} & \mathbf{M}_{43} & \mathbf{M}_{44} \end{bmatrix} \begin{Bmatrix} \mathbf{C}_{wA} \\ \mathbf{C}_{wB} \\ \mathbf{C}_{uA} \\ \mathbf{C}_{uB} \end{Bmatrix} = \begin{Bmatrix} \mathbf{F}_{wA} \\ \mathbf{F}_{wB} \\ \mathbf{F}_{uA} \\ \mathbf{F}_{uB} \end{Bmatrix} \qquad (106)$$

where

$$\mathbf{C}_{p(r)q(r)} = \{C^0_{p(r)q(r)}, C^1_{p(r)q(r)}, C^2_{p(r)q(r)}, \dots\}^T, \qquad (107)$$

$$\mathbf{F}_{p(r)q(r)} = \{F^0_{p(r)q(r)}, F^1_{p(r)q(r)}, F^2_{p(r)q(r)}, \dots\}^T, \qquad (108)$$

$$K_{rs,mm'} = \delta_{rs}(1-\delta_{0m})(1-\delta_{0m'})(\delta_{mm'} + S^{p(r)q(r)}_{rs,m'm})\lambda^4_{q(r)m'} + \delta_{rs}\delta_{0m'}\delta_{m0}S^{p(r)q(r)}_{rs,mm'}$$

$$+ (1-\delta_{rs})\delta_{m0}S^{p(r)q(r)}_{rs,mm'} \quad , \qquad (109)$$

$$M_{rs,mm'} = \frac{\rho_{q(r)}A_{q(r)}}{\overline{D}_r}\left(\delta_{rs}(1-\delta_{0m})(1-\delta_{0m'})(\delta_{mm'} + S^{p(r)q(r)}_{rs,mm'} + S^{p(r)q(r)}_{rs,m'm} + Z^{p(r)q(r)}_{rs,mm'})\right.$$

$$+ \delta_{rs}\delta_{0m'}(S^{p(r)q(r)}_{rs,mm'} + Z^{p(r)q(r)}_{rs,mm'}) + \delta_{rs}\delta_{m0}(S^{p(r)q(r)}_{rs,m'm} + Z^{p(r)q(r)}_{rs,mm'}) + \delta_{rs}\delta_{0m'}\delta_{m0}(2 + Z^{p(r)q(r)}_{rs,mm'})$$

$$+ (1-\delta_{rs})(1-\delta_{0m})(S^{p(r)q(r)}_{rs,mm'} + Z^{p(r)q(r)}_{rs,mm'}) + (1-\delta_{rs})\delta_{m0}Z^{p(r)q(r)}_{rs,mm'}\right)$$

$$\text{for } m, m' = 0, 1, 2, 3, \dots \qquad (110)$$

$$F_{p(r)q(r),m} = 2/\overline{D}_r L_{q(r)} \int_0^{L_{q(r)}} \left(\cos\lambda_{q(r)m}x + \boldsymbol{\zeta}_{p(r)}(x)^T \tilde{\mathbf{H}}_r \mathbf{Q}_{rm}\right) f_{p(r)q(r)}(x)dx \qquad (111)$$

and

$$\overline{D}_r = \begin{cases} D_{q(r)}, & \text{for } r = 1, 2 \\ E_{q(r)}A_{q(r)}, & \text{for } r = 3, 4 \end{cases} \qquad (112)$$

The matrices, S and Z in Eqs. (109) and (110) are also defined in Appendix A.

In the above derivations, each beam has been implicitly treated as uniform; that is, the cross-sectional areas, mass densities and bending rigidities are all constant over each beam. However, while this condition imposes a limit to many other techniques, it is not a problem for the current solution method since the displacement expressions given in Eqs. (80) and (81) are equally valid for a non-uniform beam. When a non-uniform beam is involved, only the governing equations and boundary/coupling conditions related to that particular beam need to be modified accordingly; but all the solution algorithms and procedures will be essentially the same. Additionally, a load is here allowed to be applied anywhere on a beam, rather than only to the ends as explicitly required in the SEM and DSM. Thus, the current method can be directly extended to non-uniform beams.

EXAMPLE 3. Consider two beams elastically coupled together at a 45° angle, as shown in Figure 5. The left end of the horizontal beam (beam A) is fixed onto the ground, and the second beam is simply supported at the right end. The stiffness values for the coupling springs are given in Figure 5. The related parameters are summarized in Table 4. Table 5 shows the first 10 natural frequencies for this two-beam system. The FEA results are also given there for comparison.

Table 4. Material and beam parameters

Properties	Beam A	Beam B
Length [m]	1.0	1.5
Young's Modulus [N/m]	2.07e+11	2.07e+11
Cross-Section Area [m^2]	5.0e-5	1.5e-5
Moment of Inertia [Kg.m^4]	1.0e-10	5.0e-11
Density [Kg/m^3]	7.8e+3	7.8e+3

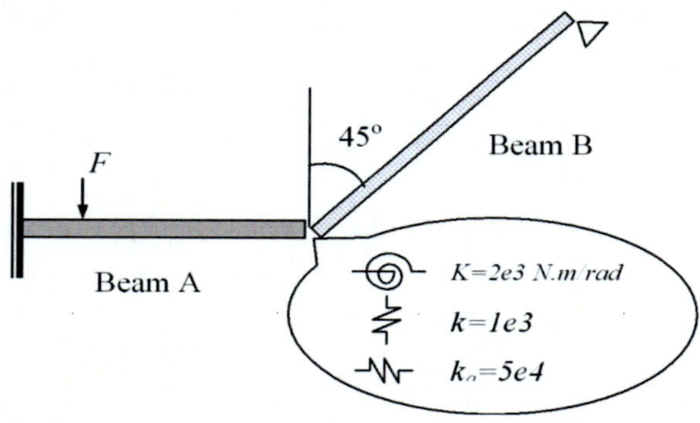

Figure 5. Two beams elastically connected together at an angle of 45 degrees.

Table 5. A comparison of the first 10 natural frequencies

Modes	Current	FEA	Difference (%)
1	8.836	8.829	0.069
2	13.398	13.344	0.408
3	28.011	27.932	0.282
4	35.402	35.422	0.057
5	61.775	61.709	0.108
6	80.624	80.663	0.049
7	107.95	107.87	0.078
8	148.29	148.31	0.013
9	164.31	164.21	0.060
10	223.56	223.23	0.147

A total of 250 beam elements of equal size, 0.01 m, have been used in the FEA (ANSYS) model, which is considered find enough to adequately capture the spatial variations of these lower order modes.

VIBRATIONS OF MULTI-SPAN BEAMS

The vibrations of multi-span beams are of considerable interest to many engineering areas such as bridge dynamics, vibration localizations and vibro-acoustic responses of periodic structures, system parameters identifications and structural health monitoring, and so on. While modern numerical methods such as the finite element methods can be generally used for the dynamic analyses of multi-span beams, some insightful information or details may be easily lost in the numerical processes. In addition, a grid-based solution method tends to become cumbersome in certain applications involving, such as, moving loads. Thus, analytical solutions are often desired in studying the dynamics of distributed systems.

Of many solution methods, the assumed mode method [28-32] is perhaps the most popular one for solving multi-span beam problems. In this method, an intermediate support is often viewed as the lumped element which is added to the main beam. The assumed modes are typically the beam functions that are usually extracted under the unloaded or unconstrained conditions. Other commonly used methods include Laplace transformations [33-36], the methods of Lagrange multipliers [37-39], the Green's function methods [40-44], and dynamic stiffness methods [45-47] to name a few. These methods will typically require a varying level of modifications or adaptations to account for the variations in boundary conditions, intermediate supports, and/or the number of spans. For instance, when the unconstrained beam functions are used as the assumed mode shapes, one typically needs to first determine the eigenfunctions for the given boundary conditions. This problem itself may become a sizeable task if the beam is elastically restrained at either or both ends. In addition, the beam eigenfunctions tend to become numerically unstable for large modal indexes, which demands special treatments in numerical calculations. In this section, the modified Fourier series method will be extended to the multi-beam systems. It will be shown that this method provides an accurate, general and unified solution to the most general kind of beam systems with respect to the arbitrariness of the boundary, coupling, and load conditions.

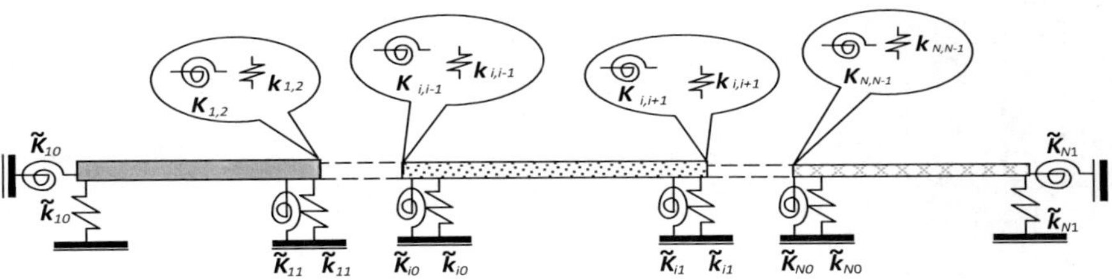

Figure 6. An illustration of a multi-span beam system.

Figure 6 illustrate a dynamic system consisting of multiple beams coupled together via a set of joints which are modeled by linear and rotational springs. The use of the coupling springs between two adjacent beams allows accounting for the effects of some non-rigid or resilient connectors. The conventional rigid connectors can be considered as a special case when the stiffnesses of these springs become substantially large with reference to the bending rigidities of the involved beams. Each of beams may also be independently supported on a set of elastic restraints at both ends.

The differential equation for the vibration of the *i-th* beam is still governed by Eq. (66). The coupling conditions between two adjacent beams can be expressed as

at $x_i = 0$,

$$k_{i,i-1}(w_i - w_{i-1}) + \tilde{k}_{i0}w_i = -D_i w_i''' , \tag{113}$$

$$K_{i,i-1}(w_i' - w_{i-1}') + \tilde{K}_{i0}w_i' = D_i w_i'' \quad ; \tag{114}$$

at $x_i = L_i$,

$$k_{i,i+1}(w_i - w_{i+1}) + \tilde{k}_{i1}w_i = D_i w_i''' , \tag{115}$$

$$K_{i,i+1}(w_i' - w_{i+1}') + \tilde{K}_{i1}w_i' = -D_i w_i'' \tag{116}$$

where $k_{i,j}$ and $K_{i,j}$ denote the stiffnesses of the linear and rotational springs at the junction of beam i and j, respectively; $\tilde{k}_{i,0}, \tilde{k}_{i,1}$ are the stiffnesses of linear springs, and $\tilde{K}_{i,0}, \tilde{K}_{i,1}$ are the stiffnesses of the rotational springs at the left and right ends of beam i, respectively.

Here the displacement on each beam can still be expanded into a Fourier series in the form of Eq. (80). As a result, the above coupling conditions can be rewritten as

$$\mathbf{H}_{i,i-1}\boldsymbol{\alpha}_{i-1} + \mathbf{H}_{i,i}\boldsymbol{\alpha}_i + \mathbf{H}_{i,i+1}\boldsymbol{\alpha}_{i+1} = \sum_{m=0}^{\infty}\left(\mathbf{Q}_{i,i-1}^m\mathbf{A}_{i-1,m} + \mathbf{Q}_{i,i}^m\mathbf{A}_{i,m} + \mathbf{Q}_{i,i+1}^m\mathbf{A}_{i+1,m}\right) \tag{117}$$

where

$$\mathbf{H}_{i,i} = \begin{bmatrix} \dfrac{8(k_{i,i-1}+\tilde{k}_{i0})L_i^3}{360}+D_i & \dfrac{7(k_{i,i-1}+\tilde{k}_{i0})L_i^3}{360} & \dfrac{-(k_{i,i-1}+\tilde{k}_{i0})L_i}{3} & \dfrac{-(k_{i,i-1}+\tilde{k}_{i0})L_i}{6} \\[4pt] \dfrac{7(k_{i,i+1}+\tilde{k}_{i1})L_i^3}{360} & \dfrac{8(k_{i,i+1}+\tilde{k}_{i1})L_i^3}{360}+D_i & \dfrac{-(k_{i,i+1}+\tilde{k}_{i1})L_i}{6} & \dfrac{-(k_{i,i+1}+\tilde{k}_{i1})L_i}{3} \\[4pt] \dfrac{L_iD_i}{3} & \dfrac{L_iD_i}{6} & (K_{i,i-1}+\tilde{K}_{i0})+\dfrac{D_i}{L_i} & \dfrac{-D_i}{L_i} \\[4pt] \dfrac{L_iD_i}{6} & \dfrac{L_iD_i}{3} & \dfrac{-D_i}{L_i} & (K_{i,i+1}+\tilde{K}_{i1})+\dfrac{D_i}{L_i} \end{bmatrix}, \tag{118}$$

$$\mathbf{H}_{i,i+1} = \begin{bmatrix} 0 & 0 & 0 & 0 \\[4pt] \dfrac{8k_{i,i+1}L_{i+1}^3}{360} & \dfrac{7k_{i,i+1}L_{i+1}^3}{360} & \dfrac{-k_{i,i+1}L_{i+1}}{3} & \dfrac{-k_{i,i+1}L_{i+1}}{6} \\[4pt] 0 & 0 & 0 & 0 \\[4pt] 0 & 0 & -K_{i,i+1} & 0 \end{bmatrix} \tag{119}$$

$$\mathbf{H}_{i,i-1} = \begin{bmatrix} \dfrac{7k_{i,i-1}L_{i-1}^{3}}{360} & \dfrac{8k_{i,i-1}L_{i-1}^{3}}{360} & \dfrac{-k_{i,i-1}L_{i-1}}{6} & \dfrac{-k_{i,i-1}L_{i-1}}{3} \\ 0 & 0 & 0 & 0 \\ 0 & 0 & 0 & -K_{i,i-1} \\ 0 & 0 & 0 & 0 \end{bmatrix} \tag{120}$$

$$\mathbf{Q}_{i,i}^{m} = \left\{ -(k_{i,i-1} + \tilde{k}_{i0}) \quad (-1)^{m}(k_{i,i+1} + \tilde{k}_{i1}) \quad -D_i \lambda_{i,m}^{2} \quad (-1)^{m} D_i \lambda_{i,m}^{2} \right\}^{\mathrm{T}} \tag{121}$$

and

$$\mathbf{Q}_{i,i-1}^{m} = \left\{ (-1)^{m} k_{i,i-1} \quad 0 \quad 0 \quad 0 \right\}^{\mathrm{T}} \text{ and } \mathbf{Q}_{i,i+1}^{m} = \left\{ 0 \quad -k_{i,i+1} \quad 0 \quad 0 \right\}^{\mathrm{T}}. \tag{122, 123}$$

Equation (117) actually represents a set of 4 linear algebraic equations that relate the 12 unknown boundary constants to the Fourier coefficients for the three involved beams. To fully determine the boundary constants, one has to apply equation (117) in turn to each beam, resulting in a total of $4N$ equations which can be expressed in matrix form as

$$\overline{\mathbf{H}}\overline{\boldsymbol{\alpha}} = \sum_{m=0}^{\infty} \overline{\mathbf{Q}}^{m} \mathbf{A}_{m} \tag{124}$$

where

$$\overline{\boldsymbol{\alpha}} = \{\boldsymbol{\alpha}_1, \boldsymbol{\alpha}_2, \ldots, \boldsymbol{\alpha}_N\}^{T}, \tag{125}$$

$$\overline{\mathbf{A}}_m = \{\mathbf{A}_{1,m}, \mathbf{A}_{2,m}, \ldots, \mathbf{A}_{N,m}\}^{T}, \tag{126}$$

$$\overline{\mathbf{H}} = \begin{bmatrix} \mathbf{H}_{1,1} & \mathbf{H}_{1,2} & 0 & \cdots & 0 & 0 & 0 & \cdots & 0 & 0 \\ \mathbf{H}_{2,1} & \mathbf{H}_{2,2} & \mathbf{H}_{2,3} & \cdots & 0 & 0 & 0 & \cdots & 0 & 0 \\ \vdots & \vdots & \vdots & \cdots & \vdots & \vdots & \vdots & \cdots & \vdots & \vdots \\ 0 & 0 & 0 & \cdots & \mathbf{H}_{i,i-1} & \mathbf{H}_{i,i-1} & \mathbf{H}_{i,i-1} & \cdots & 0 & 0 \\ \vdots & \vdots & \vdots & \cdots & \vdots & \vdots & \vdots & \cdots & \vdots & \vdots \\ 0 & 0 & 0 & \cdots & 0 & 0 & 0 & \cdots & \mathbf{H}_{N,N-1} & \mathbf{H}_{N,N} \end{bmatrix}, \tag{127}$$

and

$$
\overline{\mathbf{Q}}^m =
\begin{bmatrix}
\mathbf{Q}_{1,1}^m & \mathbf{Q}_{1,2}^m & 0 & \cdots & 0 & 0 & 0 & \cdots & 0 & 0 \\
\mathbf{Q}_{2,1}^m & \mathbf{Q}_{2,2}^m & \mathbf{Q}_{2,3}^m & \cdots & 0 & 0 & 0 & \cdots & 0 & 0 \\
\vdots & \vdots & \vdots & \cdots & \vdots & \vdots & \vdots & \cdots & \vdots & \vdots \\
0 & 0 & 0 & \cdots & \mathbf{Q}_{i,i-1}^m & \mathbf{Q}_{i,i}^m & \mathbf{Q}_{i,i+1}^m & \cdots & 0 & 0 \\
\vdots & \vdots & \vdots & \cdots & \vdots & \vdots & \vdots & \cdots & \vdots & \vdots \\
0 & 0 & 0 & \cdots & 0 & 0 & 0 & \cdots & \mathbf{Q}_{N,N-1}^m & \mathbf{Q}_{N,N}^m
\end{bmatrix}.
\tag{128}
$$

All the unknown constants can then be determined from Eq. (124) as

$$
\overline{\alpha} = \sum_{m=0}^{\infty} \widetilde{\mathbf{H}}\,\overline{\mathbf{Q}}^m \mathbf{A}_m
\tag{129}
$$

where

$$
\widetilde{\mathbf{H}} =
\left\{
\begin{array}{c}
\widetilde{\mathbf{H}}_1 \\
\vdots \\
\widetilde{\mathbf{H}}_i \\
\vdots \\
\widetilde{\mathbf{H}}_N
\end{array}
\right\}
= \overline{\mathbf{H}}^{-1}
\tag{130}
$$

and

$$
\overline{\mathbf{Q}}^m =
\left\{
\begin{array}{c}
\overline{\mathbf{Q}}_1^m \\
\overline{\mathbf{Q}}_2^m \\
\vdots \\
\overline{\mathbf{Q}}_i^m \\
\vdots \\
\overline{\mathbf{Q}}_N^m
\end{array}
\right\}
\tag{131}
$$

with

$$
\overline{\mathbf{Q}}_1^m =
\left\{
\begin{array}{c}
\mathbf{Q}_{1,1}^m \\
\mathbf{Q}_{2,1}^m \\
0 \\
0 \\
0
\end{array}
\right\},
\quad
\overline{\mathbf{Q}}_i^m =
\left\{
\begin{array}{c}
0 \\
\mathbf{Q}_{i-1,i}^m \\
\mathbf{Q}_{i,i}^m \\
\mathbf{Q}_{i+1,i}^m \\
0
\end{array}
\right\},
\quad
\overline{\mathbf{Q}}_N^m =
\left\{
\begin{array}{c}
0 \\
0 \\
0 \\
\mathbf{Q}_{N-1,N}^m \\
\mathbf{Q}_{N,N}^m
\end{array}
\right\}.
\tag{132-134}
$$

Specifically, the boundary constants for beam i can be expressed as

$$\alpha_i = \sum_{m=0}^{\infty} \widetilde{\mathbf{H}}_i \overline{\mathbf{Q}}^m \mathbf{A}_{i,m} \tag{135}$$

and the corresponding displacement as

$$w_i(x) = \sum_{m=0}^{\infty} A_{i,m}(\cos\lambda_{i,m}x + \xi_i(x)^T \widetilde{\mathbf{H}}_i \overline{\mathbf{Q}}_i^m) + \sum_{\substack{j=1,N \\ j \neq i}} \sum_{m=0}^{\infty} A_{j,m}\xi_j(x)^T \widetilde{\mathbf{H}}_i \overline{\mathbf{Q}}_j^m \tag{136}$$

By substituting Eq. (136) into Eq. (66), multiplying both sides with $\cos\lambda_{im}x + \xi_i(x)^T \widetilde{\mathbf{H}}_i \overline{\mathbf{Q}}_i^m$, and integrating the result from 0 to L_i, one is able to obtain

for $m = 1,2,3\ldots$,

$$\sum_{m'=1}^{\infty}(\delta_{mm'} + S_{i,m'm}^{ii})\lambda_{i,m'}^4 A_{i,m'} - \overline{\rho}_i\omega^2 \Big[\big(S_{i,m0}^{ii} + Z_{i,m0}^{ii}\big)A_{i,0} +$$

$$\sum_{m'=1}^{\infty}\big(\delta_{mm'} + S_{i,mm'}^{ii} + S_{i,m'm}^{ii} + Z_{i,mm'}^{ii}\big)A_{i,m'} + \sum_{\substack{j=1 \\ j \neq i}} \sum_{m'=0}^{\infty}\big(S_{i,mm'}^{ij} + Z_{i,mm'}^{ij}\big)A_{j,m'}\Big] = \bar{f}_{i,m} \tag{137}$$

for $m = 0$,

$$S_{i,00}^{ii}A_{i,0} + \sum_{\substack{j=1 \\ j \neq i}} \sum_{m'=0}^{\infty} S_{i,0m'}^{ij} A_{j,m'} -$$

$$\overline{\rho}_i\omega^2\left((2 + Z_{i,00}^{ii})A_{i,0} + \sum_{m'=1}^{\infty}\big(S_{i,m'0}^{ii} + Z_{i,0m'}^{ii}\big)A_{i,m'} + \sum_{\substack{j=1 \\ j \neq i}} \sum_{m'=0}^{\infty} Z_{i,0m'}^{ij}A_{j,m'}\right) = \bar{f}_{0,m} \tag{138}$$

Equations (137) and (138) can be combined into a matrix equation

$$\big(\overline{\mathbf{K}} - \omega^2\overline{\mathbf{M}}\big)\overline{\mathbf{A}} = \overline{\mathbf{F}} \tag{139}$$

where

$$\overline{\mathbf{K}} = \begin{bmatrix} \mathbf{K}_{11} & \mathbf{K}_{12} & \cdots & \mathbf{K}_{1,i-1} & \mathbf{K}_{1,i} & \mathbf{K}_{1,i+1} & \cdots & \mathbf{K}_{1,N-1} & \mathbf{K}_{1,N} \\ \mathbf{K}_{21} & \mathbf{K}_{22} & \cdots & \mathbf{K}_{2,i-1} & \mathbf{K}_{2,i} & \mathbf{K}_{2,i+1} & \cdots & \mathbf{K}_{2,N-1} & \mathbf{K}_{2,N} \\ \vdots & \vdots & & \vdots & \vdots & \vdots & & \vdots & \vdots \\ \mathbf{K}_{i,1} & \mathbf{K}_{i,2} & \cdots & \mathbf{K}_{i,i-1} & \mathbf{K}_{i,i} & \mathbf{K}_{i,i+1} & \cdots & \mathbf{K}_{i,N-1} & \mathbf{K}_{i,N} \\ \vdots & \vdots & & \vdots & \vdots & \vdots & & \vdots & \vdots \\ \mathbf{K}_{N,1} & \mathbf{K}_{N,2} & \cdots & \mathbf{K}_{N,i-1} & \mathbf{K}_{N,i} & \mathbf{K}_{N,i+1} & \cdots & \mathbf{K}_{N,N-1} & \mathbf{K}_{N,N} \end{bmatrix} \tag{140}$$

$$\overline{\mathbf{M}} = \begin{bmatrix} \mathbf{M}_{11} & \mathbf{M}_{12} & \cdots & \mathbf{M}_{1,i-1} & \mathbf{M}_{1,i} & \mathbf{M}_{1,i+1} & \cdots & \mathbf{M}_{1,N-1} & \mathbf{M}_{1,N} \\ \mathbf{M}_{21} & \mathbf{M}_{22} & \cdots & \mathbf{M}_{2,i-1} & \mathbf{M}_{2,i} & \mathbf{M}_{2,i+1} & \cdots & \mathbf{M}_{2,N-1} & \mathbf{M}_{2,N} \\ \vdots & \vdots & & \vdots & \vdots & \vdots & & \vdots & \vdots \\ \mathbf{M}_{i,1} & \mathbf{M}_{i,2} & \cdots & \mathbf{M}_{i,i-1} & \mathbf{M}_{i,i} & \mathbf{M}_{i,i+1} & \cdots & \mathbf{M}_{i,N-1} & \mathbf{M}_{i,N} \\ \vdots & \vdots & & \vdots & \vdots & \vdots & & \vdots & \vdots \\ \mathbf{M}_{N,1} & \mathbf{M}_{N,2} & \cdots & \mathbf{M}_{N,i-1} & \mathbf{M}_{N,i} & \mathbf{M}_{N,i+1} & \cdots & \mathbf{M}_{N,N-1} & \mathbf{M}_{N,N} \end{bmatrix}, \tag{141}$$

$$\overline{\mathbf{F}} = \begin{Bmatrix} \mathbf{F}_1 \\ \mathbf{F}_2 \\ \vdots \\ \mathbf{F}_i \\ \vdots \\ \mathbf{F}_N \end{Bmatrix} \quad \text{and} \quad \overline{\mathbf{A}} = \begin{Bmatrix} \mathbf{A}_1 \\ \mathbf{A}_2 \\ \vdots \\ \mathbf{A}_i \\ \vdots \\ \mathbf{A}_N \end{Bmatrix} \tag{142, 143}$$

with the notations $\mathbf{F}_i = \{f_{i,0}, f_{i,1}, f_{i,2}, ...\}^T$ and $\mathbf{A}_i = \{A_{i,0}, A_{i,1}, A_{i,2}, ...\}^T$.

More explicitly, the elements of the stiffness and mass matrices in Eqs. (140) and (141) are defined as follows:

for $m, m' = 0, 1, 2, 3, ...,$

$$K_{pq,mm'} = \delta_{pq}(1-\delta_{0m})(1-\delta_{0m'})(\delta_{mm'} + S^{pq}_{p,m'm})\lambda^4_{pm'}$$

$$+ \delta_{pq}\delta_{0m'}\delta_{m0}S^{pq}_{p,00} + (1-\delta_{pq})\delta_{m0}S^{pq}_{p,0m'} \tag{144}$$

$$M_{pq,mm'} = \overline{\rho}_i \Big(\delta_{pq}(1-\delta_{0m})(1-\delta_{0m'})(\delta_{mm'} + S^{pq}_{p,mm'} + S^{pq}_{p,m'm} + Z^{pq}_{p,mm'}) +$$

$$\delta_{pq}\delta_{0m'}\delta_{m0}(2 + Z^{pq}_{p,00}) + \delta_{pq}\delta_{0m'}(S^{pq}_{p,m0} + Z^{pq}_{p,m0}) +$$

$$\delta_{pq}\delta_{m0}(S^{pq}_{p,m'0} + Z^{pq}_{p,m'0})(-1)^{\delta_{m0}+\delta_{m'0}-1} +$$

$$(1-\delta_{pq})(1-\delta_{0m})(S^{pq}_{p,mm'} + Z^{pq}_{p,mm'}) + (1-\delta_{pq})\delta_{m0}Z^{pq}_{p,0m'} \Big) \quad ,$$

$$(\overline{\rho}_p = \rho_p S_p / D_p) \tag{145}$$

and

$$f_{p,m} = 2/D_p L_p \int_0^{L_p} \left(\cos\lambda_{pm}x + \xi_p(x)^T \widetilde{\mathbf{H}}_p \overline{\mathbf{Q}}^m_p \right) f_p(x) dx . \tag{146}$$

Equation (139) represents a set of linear algebraic equations from which the unknown Fourier coefficients, and hence the responses, can be readily determined for a given loading condition. EXAMPLE 4. Consider a case involving three beams elastically coupled together, as illustrated in Figure 7. The related beam parameters are summarized in Table 6. The left end of the system is clamped and the right end is simply supported. In addition, each beam is independently supported at both ends in the form of elastic constraints. The details of the supporting and coupling conditions are described in Table 7 in terms of the spring constants. For example, the simply supported condition at the right end is accomplished through setting $\tilde{k}_{3,1} = 10^{10}$ and $\tilde{K}_{3,1} = 0$ (refer to Figure 7 for the notations). A moderate stiffness value indicates the possibility of a relative motion between the two corresponding degrees of freedom (one of them may represent the ground).

Table 6. Beam parameters and material properties

Variables	Beam 1	Beam2	Beam3
L	1.0	1.5	2.0
S	5×10^{-5}	1.5×10^{-5}	5×10^{-5}
I	10^{-10}	5×10^{-11}	10^{-10}
E	2.07×10^{11}	2.07×10^{11}	2.07×10^{11}
ρ	7800	7800	7800

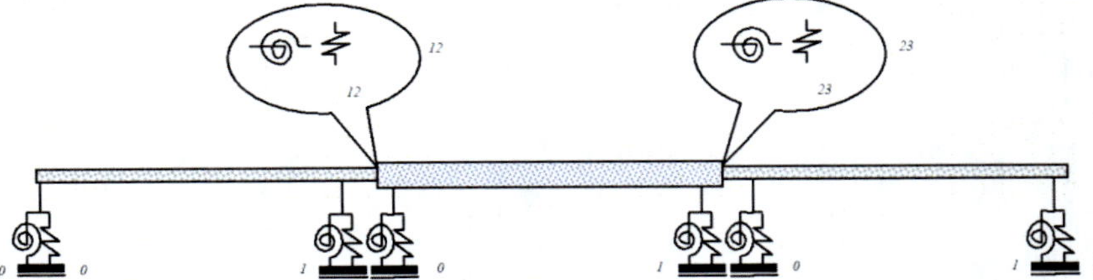

Figure 7. Three elastically coupled beams with arbitrary boundary supports.

Table 7. Stiffness values for the boundary and coupling springs in the three-beam system

Spring constants for beam 1	Spring constants for beam 2	Spring constants for beam3	Spring constants for joints
$\tilde{k}_{1,0} = 10^{10}$	$\tilde{k}_{2,0} = 4000$	$\tilde{k}_{3,0} = 5000$	$k_{1,2} = 1000$
$\tilde{k}_{1,1} = 5000$	$\tilde{k}_{2,1} = 4000$	$\tilde{k}_{3,1} = \infty$	$k_{2,3} = 1000$
$\tilde{K}_{1,0} = \infty$	$\tilde{K}_{2,0} = 1000$	$\tilde{K}_{3,0} = 2000$	$K_{1,2} = 200$
$\tilde{K}_{1,1} = 2000$	$\tilde{K}_{2,1} = 1000$	$\tilde{K}_{3,1} = 0$	$K_{2,3} = 200$

Listed in Table 8 are the ten lowest natural frequencies for this three-beam system. The FEA results obtained using an ANSYS model are also presented there as a reference. In the FEA model the element size is chosen as 0.01 m. In the current calculation, the series expansion for each beam is truncated to include only the first 10 terms (M=10).

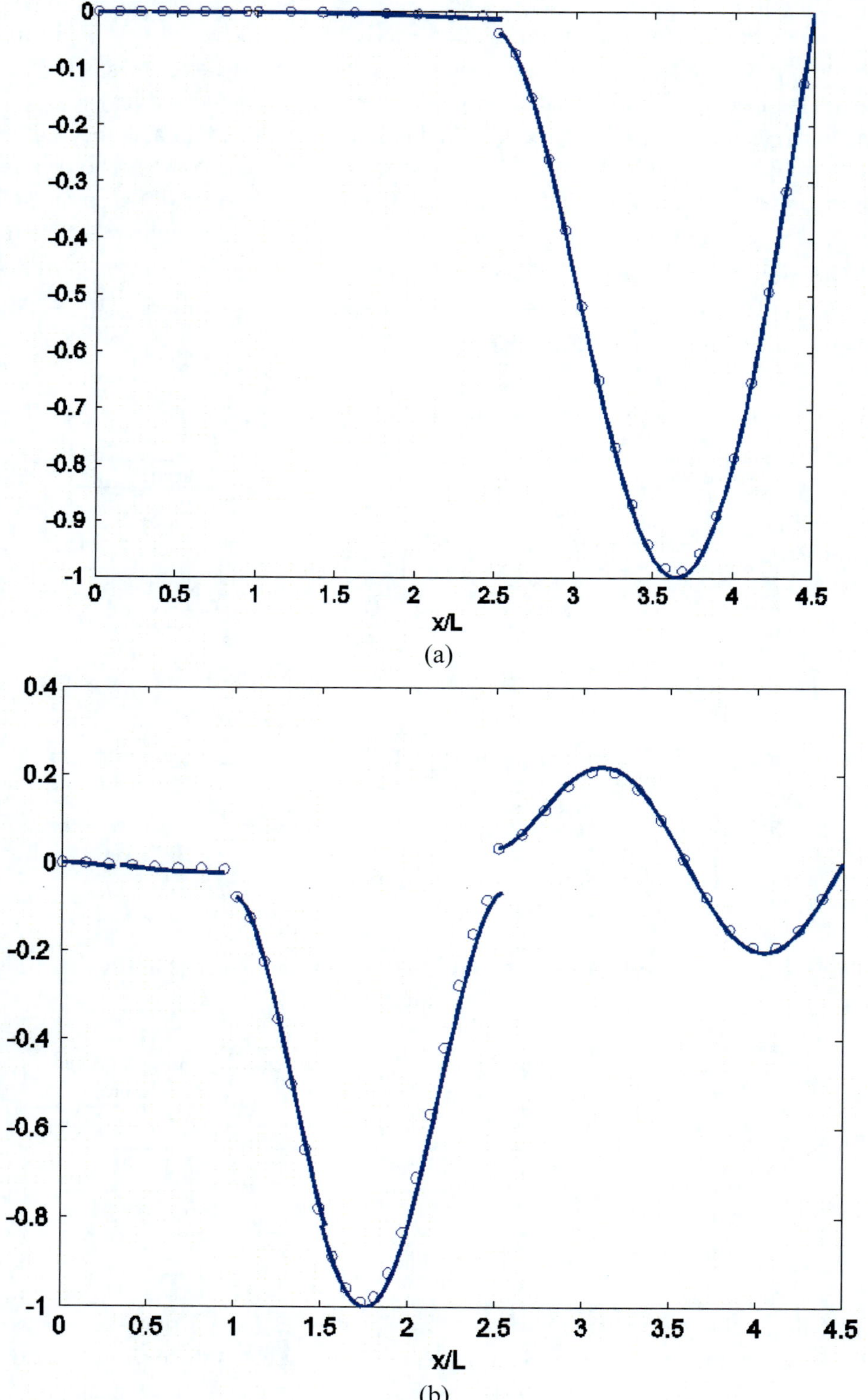

Figure 8. Selected plots of the mode shapes for the three elastically coupled beams, current method; o, FEA: (a) the first mode, and (b) the third mode.

Table 8. Ten lowest natural frequencies for the three-beam system

Mode	FEA	Current
1	4.3675	4.3675
2	13.646	13.646
3	13.842	13.848
4	21.689	21.689
5	26.697	26.697
6	34.322	34.325
7	42.115	42.116
8	46.902	46.903
9	58.147	58.149
10	62.465	62.467

The mode shapes for two "randomly" selected modes are plotted in Figure 8 together with the FEA results. Because of the elastic couplings, the displacements on the beams are no longer continuous across the junctions. It is seen (and also predictable) that the displacement jumps at the junctions tend to increase with the modal order. These plots also show the presence of mode or vibration localization. Traditionally, the vibration localization is associated with a slight dislocation of an immediate support of otherwise perfectly periodic multi-span beam. It should become clear that the coupling conditions add another dimension to this phenomenon, and potentially play an important role in influencing or controlling the vibration localization.

In many engineering applications, such as bridge dynamics, it is necessary to consider traveling loads. In such a case, Eq. (66) needs to be modified to

$$D_i \, \partial^4 w_i(x,t)/\partial x^4 - \rho_i A_i \, \partial^2 w_i(x,t)/\partial t^2 = \sum_{j=1}^{J} F_j \delta(x - x_{j,i}^f(t)) \quad (i = 1, 2, ..., N), \quad (147)$$

where J is the number of loads acting on the i-*th* beam at t moment, F_j is the magnitude of the j-*th* load, δ is the Dirac delta function, and $x_{j,i}^f(t)$ is the j-*th* load position measured from the left end of the i-*th* beam. Sometimes, unit step functions are used in Eq. (147) to explicitly specify whether a moving load is present to a span or not. For simplicity, they are not included here since the load position described in terms of the delta function has already contained this information, that is, the j-*th* load is not present to the i-*th* beam if $x_{j,i}^f(t) > L_i$ or < 0.

When a single load travels along a beam, the dynamic response can be treated as having two parts: the forced vibration caused by the load directly acting on the beam, and the residual free vibration caused by the load that has passed the beam [48]. As a consequence, phenomena of resonance and cancellation may occur when a bridge is under the action of multiple loads [48, 49]. Therefore, the loading conditions (e.g., the number of loads and their travelling speeds) are of critical importance to a bridge design. From mathematical point of view, however, Eq. (147) simply represents a linear system whose response to multiple loads can be actually considered as the superposition of its responses to each individual load.

The load profile is often defined by

$$\ddot{x}_f(t) = a = \text{const.}, \quad \dot{x}_f(t) = v = v_0 + at, \text{ and } x_f(t) = v_0 t + at^2/2,$$

where $x_f(t)$ is the load position measured from the left end of the first beam, a is the acceleration of the moving load, $v = v(t)$ is its velocity, and v_0 is the initial velocity at time $t = 0$ when the load is just about to enter the first beam.

EXAMPLE 5. In order to validate the current model, we will consider a simple multi-span beam problem that was previously studied in [50]. As illustrated in Figure 9, this problem involves a three-span stepped beam subjected to a single concentrated moving load. The relevant beam and material parameters are listed in Table 9. Under the current framework, this stepped continuous beam can be viewed as a collection of three separate beams that are rigidly coupled together. The continuous beam is assumed to be simply supported at its ends and at the two joint locations.

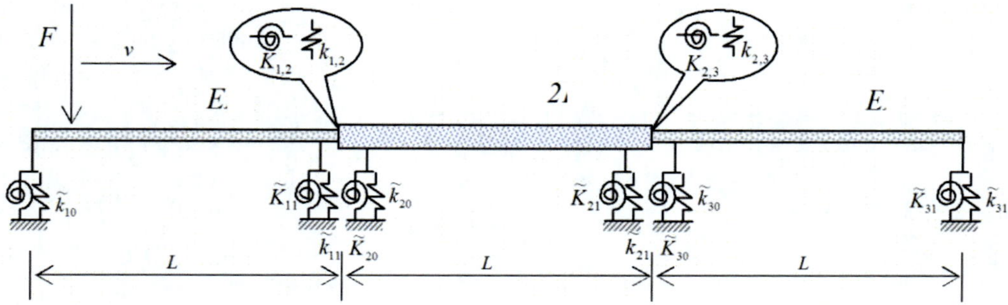

Figure 9. A three-span beam with non-uniform cross-section under a moving load.

Table 9. Beam and material properties

Parameters	Values
L	$20\,(m)$
ρ	$7800\,(Kg/m^3)$
ρA	$1000\,(Kg/m)$
EI	$1.96 \times 10^9\,(Nm^2)$
E	$10.48 \times 10^{10}\,(N/m^2)$
F	$9.48 \times 10^3\,(N)$

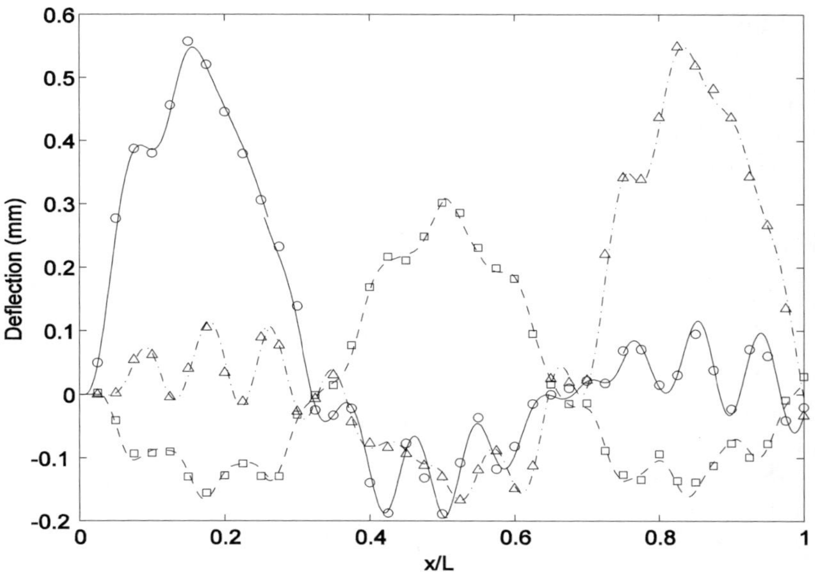

Span 1: —, Current, , [50]; Span 2: ---, Current, , [50]; Span 3: -·-, Current, Δ, [50].

Figure 10. Flexural deflection at the midpoint of each span for $v = 35.57 m/s$.

Assume the beam is subjected to a point load, $F = 9.48 \times 10^3 N$, moving at a constant speed $v = 35.57 m/s$. Plotted in Figure 10 are the corresponding deflections at the midpoint locations of all three spans. The results obtained by Henchi and Fafard [50] are also shown there for comparison. This problem was also studied by Dugush and Eisenberger [51] for different beam parameters and load profiles. It suffices to say that the current results also match closely with those given in [51].

EXTENDING THE MODIFIED FOURIER SERIES METHOD TO 2-D STRUCTURAL COMPONENTS - PLATES WITH ARBITRARY BOUNDARY CONDITIONS

Transverse vibrations of rectangular plates with various boundary conditions have been extensively studied in the literature, as comprehensively reviewed by Leissa [1]. While exact solutions are available for plates which are simply supported along at least one pair of opposite edges, it has been widely accepted that there exists no exact solution for other more general boundary conditions. Accordingly, a variety of approximate or numerical solution techniques have been employed to solve plate problems under different boundary conditions. Rayleigh-Ritz method is one of the most widely used techniques for obtaining an approximate solution. When the Rayleigh-Ritz method is employed in solving plate problems, the displacement function is often expressed in terms of characteristic functions obtained for beams with similar boundary conditions [52-56]. However, even only considering the four simplest homogeneous cases (i.e., simply supported, clamped, free and guided), one shall realize that they can be combined into 55 different types of boundary conditions for a rectangular plate. Thus, the use of beam functions will lead to at least a very tedious solution process, not to mention the potential difficulties associated with the numerical instability of the higher order characteristic functions. In addition, expressing a solution as an expansion of beam functions will clearly becomes impossible when the boundary condition becomes non-uniform along an edge such as varying elastic restraints, point supports, partial supports, and their combinations.

For a rectangular plate of length a and width b, its displacement function can be represented by a Fourier series expansion of the form [57]

$$w(x,y) = \sum_{m=0}^{\infty}\sum_{n=0}^{\infty} A_{mn} \cos\lambda_{am}x \cos\lambda_{bn}y + \sum_{l=1}^{4}\left(\xi_b^l(y)\sum_{m=0}^{\infty} c_m^l \cos\lambda_{am}x + \xi_a^l(x)\sum_{n=0}^{\infty} d_n^l \cos\lambda_{bn}y \right) \tag{148}$$

where $\lambda_{am} = m\pi/a$, $\lambda_{bn} = n\pi/b$, and $\xi_a^l(x)$ (or $\xi_b^l(y)$) represent a set of sufficiently smooth closed-form functions defined over $[0, a]$ (or $[0, b]$). In the previous discussions, they are specifically selected as the polynomials. To show the arbitrariness in selecting these auxiliary functions, they are here alternatively chosen as trigonometric functions to take advantage of their "invariance" under differential operations, that is,

$$\xi_a^1(x) = \frac{9a}{4\pi}\sin\left(\frac{\pi x}{2a}\right) - \frac{a}{12\pi}\sin\left(\frac{3\pi x}{2a}\right), \tag{149}$$

$$\xi_a^2(x) = -\frac{9a}{4\pi}\cos\left(\frac{\pi x}{2a}\right) - \frac{a}{12\pi}\cos\left(\frac{3\pi x}{2a}\right), \tag{150}$$

$$\xi_a^3(x) = \frac{a^3}{\pi^3}\sin\left(\frac{\pi x}{2a}\right) - \frac{a^3}{3\pi^3}\sin\left(\frac{3\pi x}{2a}\right), \tag{151}$$

and

$$\xi_a^4(x) = -\frac{a^3}{\pi^3}\cos\left(\frac{\pi x}{2a}\right) - \frac{a^3}{3\pi^3}\cos\left(\frac{3\pi x}{2a}\right). \tag{152}$$

It is easy to verify that $\xi_a^{1\prime}(0) = \xi_a^{2\prime}(a) = \xi_a^{3\prime\prime\prime}(0) = \xi_a^{4\prime\prime\prime}(a) = 1$, and all the other first and third derivatives are identically equal to zero along the edges. These conditions are not necessary, but make it easier to understand the meanings of the 1-D Fourier series expansions: each of them represents either the first or the third derivative of the displacement function at one of the edges.

It can be proven mathematically that the series expansion given in Eq. (148) is able to expand and uniformly converge to *any* function $f(x, y) \in C^3$ for $\forall(x, y) \in D$: ([0, a]⊗[0, b]). Also, this series can be simply differentiated, term-by-term, to obtain the uniformly convergent series expansions for up to the fourth-order derivatives. Mathematically, an exact displacement (or classical) solution is a *particular* function $w(x, y) \in C^3$ for $\forall(x, y) \in D$ which satisfies the governing equation at *every* field point and the boundary conditions at *every* boundary point. Thus, the remaining task for seeking an exact displacement solution will simply involve finding a set of expansion coefficients to ensure the governing equation and the boundary conditions to be satisfied by the current series solution *exactly on a point-wise basis*. In reality, since the series expansion has to be truncated to a finite number of terms in numerical calculations, the term "satisfied exactly" should be understood as *satisfied to any specified accuracy*.

EXAMPLE 6. Consider a plate which is clamped at $x=0$ and free at all other edges (C-F-F-F). The free-edge condition is easily created by setting the stiffness constants for both springs equal to zero. The first six frequency parameters are presented in Table 10. The upper bound for frequency parameter was calculated using Rayleigh-Ritz method by taking the first 50 admissible products of beam functions [1]. The FEM results were obtained from a model which consists of 300 linear rectangular elements along each edge.

Table 10. Frequency parameters, $\Omega = \omega a^2 \sqrt{\rho h / D}$, for C-F-F-F rectangular plate with different aspect ratios

$r=a/b$	$\Omega = \omega a^2 \sqrt{\rho h / D}$					
	1	2	3	4	5	6
1.0	3.470	8.504	21.279	27.201	30.948	54.185
	3.430*		20.874	26.501		51.502
	3.482**		21.367	27.278		54.301
	3.471[†]	8.506	21.283	27.199	30.953	54.179
1.5	3.453	11.654	21.463	39.319	53.544	61.606
2.0	3.439	14.800	21.430	48.171	60.143	92.507
2.5	3.427	17.96	21.392	57.209	60.115	105.92
	3.428[†]	17.962	21.397	57.216	60.130	105.93
3.0	3.418	21.135	21.358	60.010	66.365	118.02
3.5	3.411	21.328	24.320	59.905	75.611	117.86
4.0	3.405	21.302	27.513	59.810	84.929	117.67

* Results from reference [1] on page 80 for lower bounds.
** Results from reference [1] on page 80 for upper bounds.
[†] Results from FEM with 300 x 300 elements.

Table 11. Frequency parameters, $\Omega = \omega a^2 \sqrt{\rho h / D}$, for S-S-S-S square plate with uniform rotational restraint along edges

$K \cdot a / D$	$\Omega = \omega a^2 \sqrt{\rho h / D}$					
	1	2	3	4	5	6
1	21.500	51.187	51.187	80.816	100.58	100.59
	21.496[†]	51.184	51.184	80.818	100.58	100.58
10	28.501	60.215	60.215	90.808	111.19	111.41
	28.50*	60.22	60.22	90.81	111.2	111.4
	28.489[†]	60.196	60.196	90.790	111.16	111.39
100	34.671	70.780	70.780	104.45	127.02	127.61
	34.67*	70.78	70.78	104.5	127.0	127.6
	34.668[†]	70.771	70.771	104.44	127.01	127.59
1000	35.842	73.103	73.103	107.79	131.06	131.68
	35.842[†]	73.100	73.100	107.78	131.06	131.68

* Results from reference [58].
[†] Results from FEM with 300 x 300 elements.

EXAMPLE 7. This problem involves a simply supported square plate with a uniform elastic restraint against rotation along each edge. The calculated frequency parameters are given in Table 11 together with those previously obtained using different models. These three sets of results match very well with each other.

This modified Fourier series method has also been successfully extended to the vibrations of plates with varying boundary supports [59], and structural analyses of composite plates [60].

DYNAMIC ANALYSIS OF COMPLEX STRUCTURES

In the above discussions, the solutions have been sought in strong form by letting the displacement expressions to simultaneously satisfy both the governing and boundary/coupling conditions exactly on a point-wise basis. While such a solution strategy can still be applied to more complicated structures, the resulting system stiffness and mass matrices will generally become fully populated which is not preferred numerically. Instead, a weak form of solutions is typically more suitable for the dynamic analysis of complex structures. To illustrate the procedures, take the simple 2-D frame in Figure 11 for example. In order to be able to account for a wide variety of the possible coupling configurations, a junction is considered as a collection of a number of joints with each of them being uniquely used to connect only a pair of beams.

For example, if three beams meet at a junction, then three independent joints will be specified there (between beam 1 and 2, beam 1 and 3, and beam 2 and 3). Each joint will be represented by three springs respectively corresponding to the axial, flexural and rotational displacements. When the beams are rigidly connected together, we need to simply set the spring constants to be infinitely large.

The flexure and axial displacements for each beam, e.g., the ith beam, will be sought in the form of

$$w_i = \sum_{m=0}^{\infty} A_{iw,m} \cos\left(\frac{m\pi}{L_i}\right) + \sum_{n=1}^{4} B_{iw,n} \sin\left(\frac{n\pi}{L_i}\right) \tag{153}$$

$$u_i = \sum_{m=0}^{\infty} A_{iu,m} \cos\left(\frac{m\pi}{L_i}\right) + \sum_{n=1}^{2} B_{iu,n} \sin\left(\frac{n\pi}{L_i}\right) \qquad 0 \le x \le L_i \tag{154}$$

where L_i is the length of the ith beam, and $A_{i\alpha}$ and $B_{i\alpha}$ ($\alpha=w$ or u) are the Fourier coefficients to be determined. In comparison with being polynomials in Eq. (20), the auxiliary functions are here chosen as sine functions to show the flexibility in defining them.

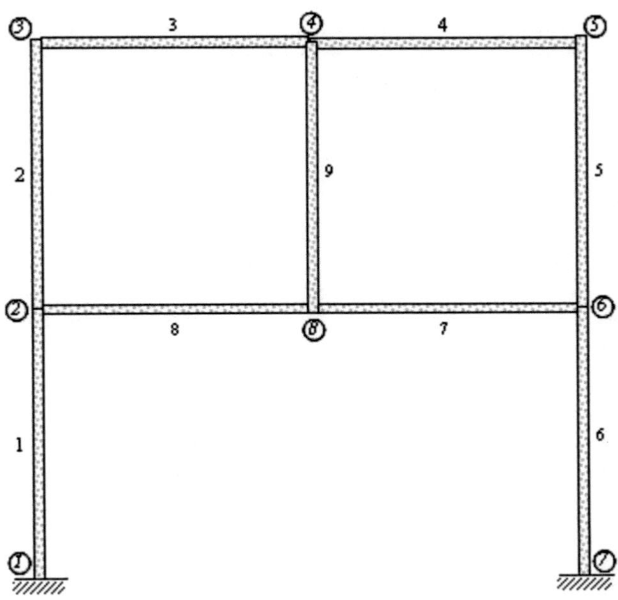

Figure 11. An exemplary 2-D frame structure.

In seeking a strong form of solution, the Fourier coefficients are solved by directly using both the governing equations and the boundary/coupling conditions. Consequently, only the coefficients A_{ia} are actually considered as a set of independent variables to be solved from the governing differential equations. As an alternative solution process which is probably preferred in modeling complex structures, all the expansion coefficients are now treated as independent variables, and solved, in an approximate manner, by using the powerful energy principles such as the Rayleigh-Ritz technique. In this way, only the neighboring structural components will directly coupled together in the final system equations, as manifested in the highly sparse stiffness and mass matrices. One of the most important outcomes is that this solution process will allow us to develop a new numerical method for complex structures by taking advantages of many of the solution algorithms and implementation schemes previously derived in the finite element methods. However, an important distinction will have to be made between the current Fourier series and the finite element methods, that is, the elements here can now represent much larger structural components of direct design concerns.

The Lagrangian L for the whole structure can be expressed as

$$L = V - T = \sum_{i=1}^{N} V_i - \sum_{i=1}^{N} T_i \tag{155}$$

where V_i and T_i are the potential and kinetic energy for ith beam, respectively.

Let k_{ij}, k_{aij} and K_{ij} denote the flexural, axial and rotational stiffnesses for the joint which connects beam i and j. The total potential energy for beam i can be calculated from

$$V_i = \frac{1}{2}\int_0^{L_i} D_i w''_i(x)^2 dx + \frac{1}{2}\int_0^{L_i} E_i A_i u'_i(x)^2 dx +$$

$$+\frac{1}{2}(1-\delta_{iN_c})\sum_{c=1}^{2}\sum_{j=i+1}^{N_c} k_{ij}\left[w_i(\zeta)\sin\frac{\alpha_{ij}}{2} + w_j(\eta)\sin\frac{\alpha_{ij}}{2} - u_i(\zeta)\cos\frac{\alpha_{ij}}{2} + u_j(\eta)\cos\frac{\alpha_{ij}}{2}\right]^2 +$$

$$+\frac{1}{2}(1-\delta_{iN_c})\sum_{c=1}^{2}\sum_{j=i+1}^{N_c} k_{aij}\left[w_i(\zeta)\cos\frac{\alpha_{ij}}{2} - w_j(\eta)\cos\frac{\alpha_{ij}}{2} + u_i(\zeta)\sin\frac{\alpha_{ij}}{2} + u_j(\eta)\sin\frac{\alpha_{ij}}{2}\right]^2 +$$

$$+\frac{1}{2}(1-\delta_{iN_c})\sum_{c=1}^{2}\sum_{j=i+1}^{N_c} K_{ij}\left[w'_i(\zeta) - w'_j(\eta)\right]^2 , \quad (\zeta = 0, L_i, \ \eta = 0, L_j) \qquad (156)$$

where α_{ij} is the angle between beam i and j, N_c is the number of beams at the junction, and $D_i = E_i I_i$ is the bending rigidity of beam i.

The total kinetic energy for beam i is defined as

$$T_i = \int_0^{L_i} \rho_i A_i \dot{w}_i(x)^2 dx + \int_0^{L_i} \rho_i A_i \dot{u}_i(x)^2 dx \qquad (157)$$

The remaining procedures of deriving the final system are well understood in numerical methods, and will no longer be repeated here for conciseness. Instead, we will proceed to presenting some numerical results.

EXAMPLE 8. The structure in Figure 11 can be naturally divided into 9 beams. Suppose that all these beams are identical: $L= 1$ m, $S=10^{-5}$ m^2, $I=1\times 10^{-10}$ m^4, $E=2.07\times 10^{11}$ N/m^2, and $\rho =7800$ kg/m^3. For simplicity, all these beams are assumed to be rigidly coupled together at each junction.

The two legs of the structure are rigidly attached to the ground. The calculated first 15 natural frequencies are listed in Table 11. The FEM results obtained from an ANSYS model are also presented as a comparison. An excellent agreement is observed between these two sets of solutions. The mode shapes for the first 4 modes are plotted in Figure 12.

Now, a unit force is applied to beam 9 in the transverse direction at a distance of $0.1L$ from joint 4 (refer to Figure 11). A uniform structural damping of 1% is applied to the whole structure.

No extra damping is applied to the joints although the damping coefficients can be separately specified for each joint. The power flows in the frame at 100 Hz is shown in Figure 13 where the arrows are used to indicate the directions and amounts of the net power flows between the beams.

Table 11. The calculated natural frequencies for the frame

Mode	FEA (Hz)	Current (Hz)
1	3.1961	3.1950
2	9.4586	9.4588
3	11.396	11.397
4	27.144	27.145
5	34.105	34.107
6	37.885	37.887
7	46.275	46.280
8	50.893	50.898
9	57.962	57.967
10	57.989	57.994
11	62.724	62.729
12	63.034	63.041
13	107.42	107.43
14	122.70	122.73
15	126.20	126.23

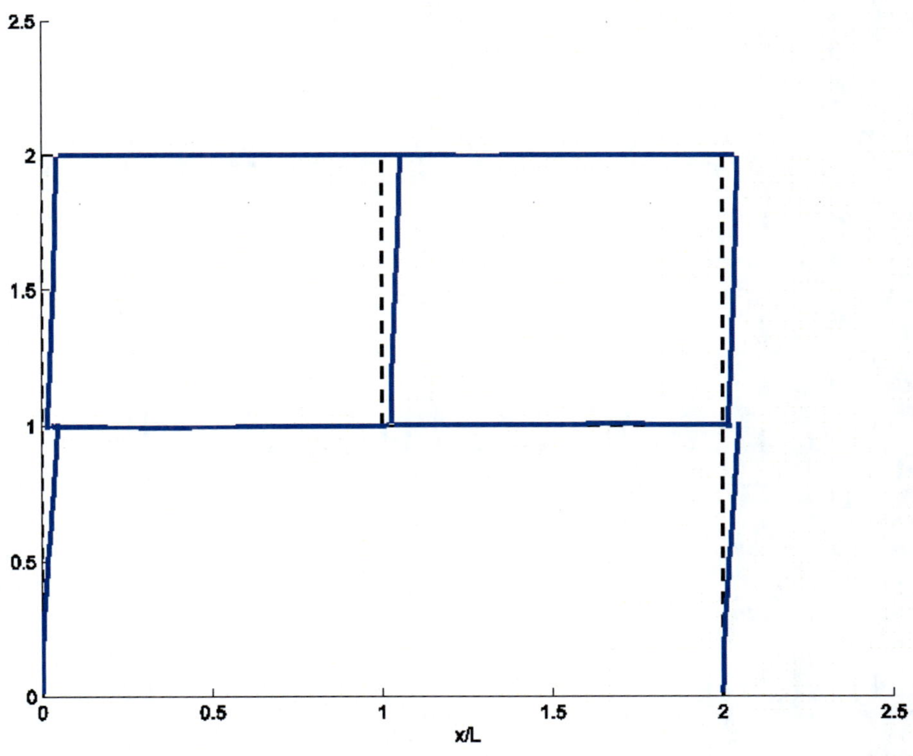

Figure 12 (Continued)

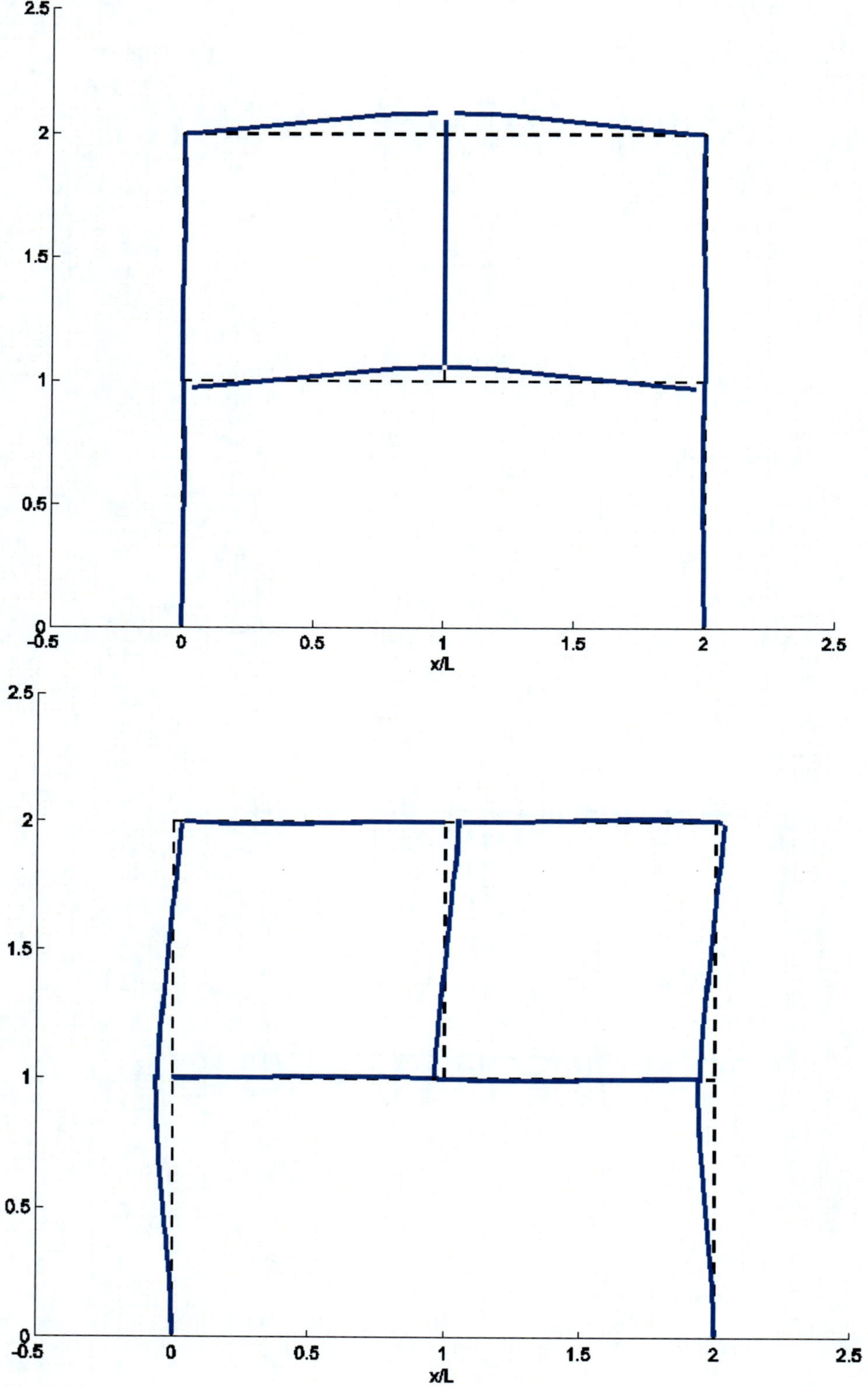

Figure 12 (Continued)

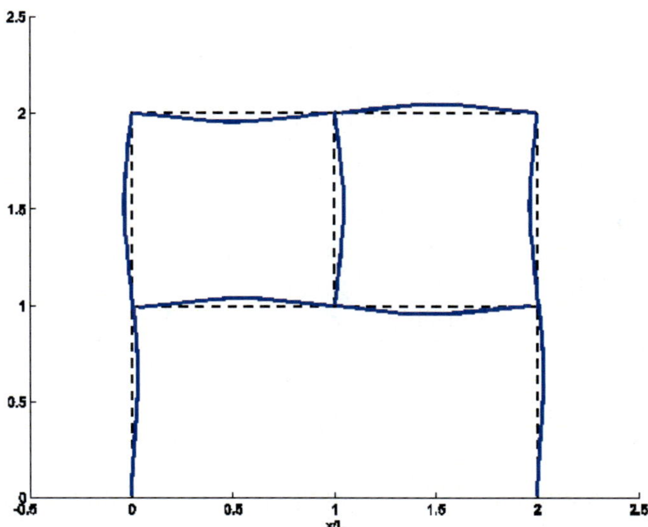

Figure 12. The mode shapes: (a) 1st mode; (b) 2nd mode; (c) 3rd mode; and (d) 4th mode.

The net power transferred from one beam to another is the sum of power flows separately associated with the flexural, rational and axial displacements. For non-dissipative joints, the total powers flowing into a junction shall be equal to zero. This power balance condition is evident from Figure 13. In comparison with vibration or strain energy levels, a power flow map usually provides more reliable information on where the vibration sources are located.

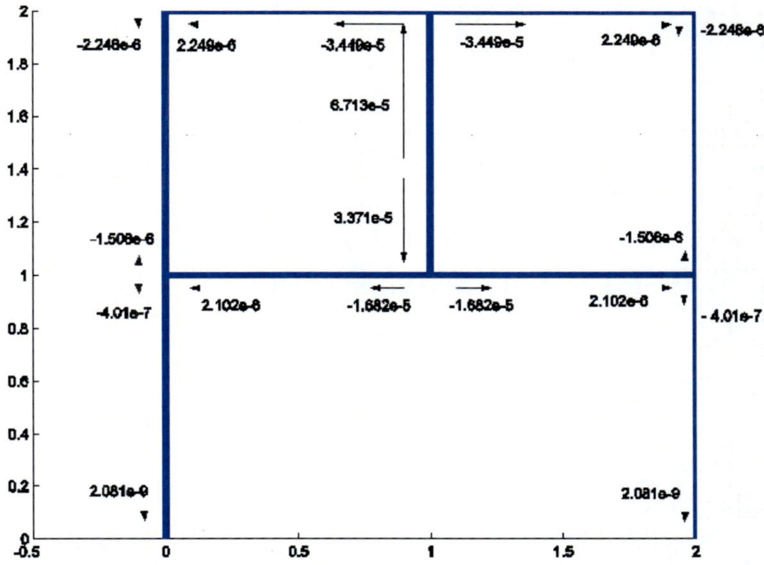

Figure 13. An illustration of power flows at 100Hz.

EXAMPLE 9. Let's now consider a more complicated structure in Figure 14 (a). As shown in Figure 14 (b), it can be naturally decomposed into 16 3-D beams and 1 plate. Three different sets of geometrical parameters are actually specified for these beams. Since this example is only intended to show the validity of the modified Fourier series method for complicated structures, a full account of this built-up structure will not be given here for compactness. Instead, it suffices to say that the modal properties can be accurately predicted as evident from the results in Table 13. A few representative modes are plotted in Figure 15. In structural dynamic analyses, the whole frequency range is often divided into three regions: low, medium, and high. At low frequencies the vibration of a structure can be satisfactorily predicted by using a deterministic model. However, as frequency increases, the dynamic characteristics of a structure typically become very sensitive to (the errors with) model variables so that a prediction based on a particular set of input data is mostly likely to be unreliable, at least, since engineering and manufacturing errors or variances always exist in reality.

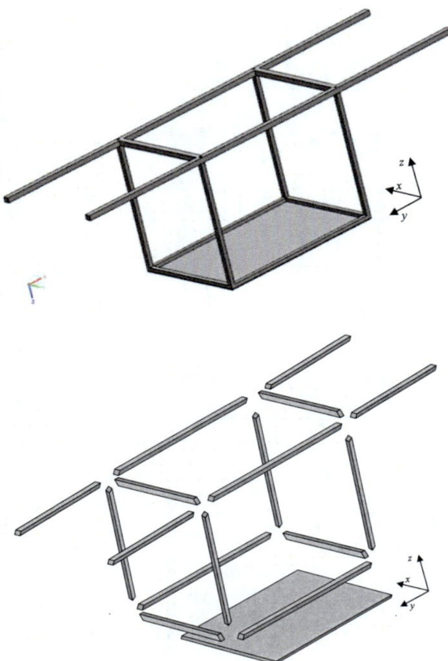

Figure 14. (a) A structure consisting of a plate and 3-D beams; (a) Its decomposition into a number of structural components.

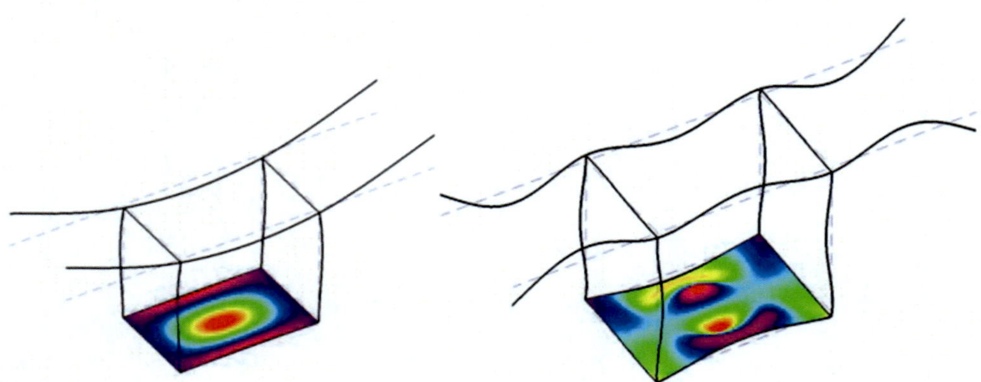

Figure 15 (Continued)

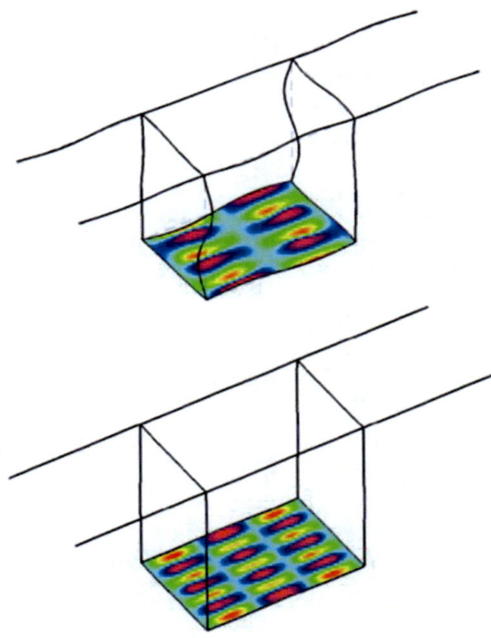

Figure 15. Mode shapes for the (a) 1st mode, 39 Hz ; (b) 51st, 429 Hz; (c) 78th mode, 782 Hz ; (d) 139th, 1552 Hz.

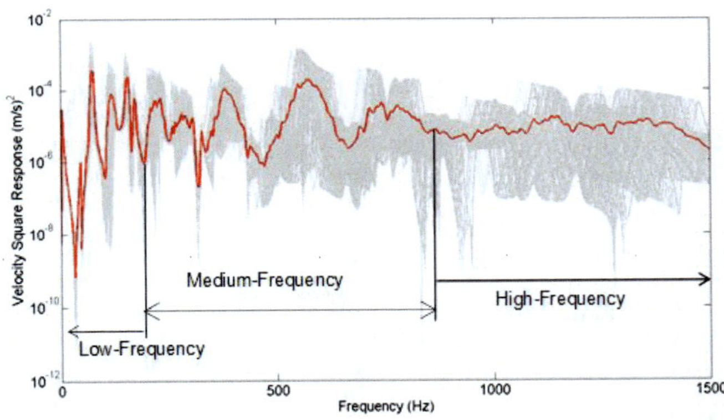

Figure 16. Vibrational responses at the same point on the plate when the plate thickness randomly fluctuates about its mean.

This can be understood from Figure 16 where the vibrations at the same point on the plate are plotted as the results of a white point force applied to the plate. The thin gray curves represent the vibrational responses corresponding to 100 different plate thickness values randomly generated by assuming they follow a zero-mean Gaussian distribution with a standard deviation 0.05. It is seen that in the low frequency region the responses can be reasonably well predicted by using the "exact" or mean thickness. At higher frequencies, however, any individual response deterministically obtained from a given thickness value tends to become practically meaningless. Instead, we will have to look at the responses from a statistical point of view. For example, the mean-square vibration represented by the thick curve in Figure 16 is a more meaningful measure of the vibration level at that point. While the mean square response curve tends to become flat at high frequencies, it does display significant fluctuations in the medium frequency region. In other words, a dynamic system can statistically exhibit a strong resonance-like behavior in the medium frequency region. As a matter of fact, the medium frequency region is often of particular interest in many noise and vibration control problems.

Table 13. The predicted natural frequencies

Modes	FEA (ANSYS)	FSEM	Relative Error (%)
1	38.9300	38.9050	0.064
2	43.3050	43.3156	0.024
3	43.9070	43.8664	0.092
4	44.1090	44.0812	0.061
5	44.2500	44.2452	0.010
6	44.7340	44.7329	0.002
7	50.7260	50.7130	0.028
8	57.8840	57.8829	0.002
9	71.7400	71.7080	0.044
10	71.8200	71.8167	0.005

CONCLUSION

A modified Fourier series method has been described as a general and effective means for determining the vibrations and dynamic characteristics of various structural components and systems. Since the displacement functions are invariably expressed as Fourier series expansions, this method can be universally applies to beams, plates and shells with different boundary conditions. Mathematically, because the improved Fourier series can expand and uniformly converge to any function with a desired smoothness, an exact (or classical) solution to a given boundary value problem can be systematically derived by letting the series simultaneously satisfy both the governing differential equations and the boundary conditions on a point-wise basis. Thus, many boundary values problems such as the vibrations of plates with general boundary conditions which are widely believed not amendable to an exact analytical solution can actually be solved under the current framework. More importantly, this modified Fourier series method can be easily extended to more complicated built-up structures. Because the modal properties for a sub-structure are no longer required, this solution method has effectively avoided many problems and difficulties related to the existing sub-structural techniques.

Finally, the much smaller final system, the mesh-less representation of a complex structure, and the continuous analytical solution over each structural component make the current method particularly attractive and useful in dealing with vibrations of complex structures in medium and high frequency regions where statistical behaviors and model uncertainties are of key concerns.

APPENDIX A: EXPRESSIONS OF H, Q, S AND Z MATRICES

The sub-matrices of the H matrix in Eq. (101) are defined as follows:

$$\mathbf{H}_{11} = \begin{bmatrix} \dfrac{8k_0 L_A^{\ 3}}{360} + D_A & \dfrac{7k_0 L^3}{360} & \dfrac{-k_0 L_A}{3} & \dfrac{-k_0 L_A}{6} \\[2mm] \dfrac{7k L_A^{\ 3}}{360} & \dfrac{8k L_A^{\ 3}}{360} + D_A & \dfrac{-k L_A}{6} & \dfrac{-k L_A}{3} \\[2mm] \dfrac{L_A D_A}{3} & \dfrac{L_A D_A}{6} & K_0 + \dfrac{D_A}{L_A} & \dfrac{-D_A}{L_A} \\[2mm] \dfrac{L_A D_A}{6} & \dfrac{L_A D_A}{3} & \dfrac{-D_A}{L_A} & K + \dfrac{D_A}{L_A} \end{bmatrix}, \tag{A1}$$

$$\mathbf{H}_{12} = \begin{bmatrix} 0 & 0 & 0 & 0 \\[2mm] \dfrac{8k L_B^{\ 3}\cos\theta}{360} & \dfrac{7k L_B^{\ 3}\cos\theta}{360} & \dfrac{-k L_B \cos\theta}{3} & \dfrac{-k L_B \cos\theta}{6} \\[2mm] 0 & 0 & 0 & 0 \\[2mm] 0 & 0 & -K & 0 \end{bmatrix}, \tag{A2}$$

$$\mathbf{H}_{14} = \begin{bmatrix} 0 & 0 \\[2mm] \dfrac{-k L_B \sin\theta}{3} & \dfrac{-k L_B \sin\theta}{6} \\[2mm] 0 & 0 \\[2mm] 0 & 0 \end{bmatrix}, \tag{A3}$$

$$\mathbf{H}_{21} = \begin{bmatrix} \dfrac{7k L_A^{\ 3}\cos^3\theta}{360} & \dfrac{8k L_A^{\ 3}\cos^3\theta}{360} & \dfrac{-k L_A \cos^3\theta}{6} & \dfrac{-k L_A \cos^3\theta}{3} \\[2mm] 0 & 0 & 0 & 0 \\[2mm] 0 & 0 & 0 & -K \\[2mm] 0 & 0 & 0 & 0 \end{bmatrix}, \tag{A4}$$

$$\mathbf{H}_{22} = \begin{bmatrix} \dfrac{8\bar{k} L_B^{\ 3}}{360} + D_B & \dfrac{7\bar{k} L_B^{\ 3}}{360} & \dfrac{-\bar{k} L_B}{3} & \dfrac{-\bar{k} L_B}{6} \\[2mm] \dfrac{7k_1 L_B^{\ 3}}{360} & \dfrac{8k_1 L_B^{\ 3}}{360} + D_B & \dfrac{-k_1 L_B}{6} & \dfrac{-k_1 L_B}{3} \\[2mm] \dfrac{L_B D_B}{3} & \dfrac{L_B D_B}{6} & K + \dfrac{D_B}{L_B} & \dfrac{-D_B}{L_B} \\[2mm] \dfrac{L_B D_B}{6} & \dfrac{L_B D_B}{3} & \dfrac{-D_B}{L_B} & K_1 + \dfrac{D_B}{L_B} \end{bmatrix}$$

$$(\bar{k} = k\cos^2\theta + k_a \sin^2\theta) \tag{A5}$$

Wen L. Li and Hongan Xu

$$\mathbf{H}_{23} = \begin{bmatrix} \dfrac{k_a L_A \sin^3\theta}{6} & \dfrac{k_a L_A \sin^3\theta}{3} \\ 0 & 0 \\ 0 & 0 \\ 0 & 0 \end{bmatrix}, \tag{A6}$$

$$\mathbf{H}_{32} = \begin{bmatrix} 0 & 0 & 0 & 0 \\ \dfrac{8k_a L_B^3 \sin\theta}{360} & \dfrac{7k_a L_B^3 \sin\theta}{360} & \dfrac{-k_a L_B \sin\theta}{3} & \dfrac{-k_a L_B \sin\theta}{6} \end{bmatrix}, \tag{A7}$$

$$\mathbf{H}_{33} = \begin{bmatrix} \dfrac{k_{a0} L_a}{3} + E_a S_a & \dfrac{k_{a0} L_a}{6} \\ \dfrac{k_a L_a}{6} & \dfrac{k_a L_a}{3} + E_a S_a \end{bmatrix} \tag{A8}$$

$$\mathbf{H}_{34} = \begin{bmatrix} 0 & 0 \\ \dfrac{k_a L_b \cos\theta}{3} & \dfrac{k_a L_b \cos\theta}{6} \end{bmatrix}, \tag{A9}$$

$$\mathbf{H}_{41} = \begin{bmatrix} \dfrac{-7kL_A^3 \sin^3\theta}{360} & \dfrac{-8kL_A^3 \sin^3\theta}{360} & \dfrac{kL_A \sin^3\theta}{6} & \dfrac{kL_A \sin^3\theta}{3} \\ 0 & 0 & 0 & 0 \end{bmatrix}, \tag{A10}$$

$$\mathbf{H}_{43} = \begin{bmatrix} \dfrac{k_a L_a \cos^3\theta}{6} & \dfrac{k_a L_a \cos^3\theta}{3} \\ 0 & 0 \end{bmatrix}, \tag{A11}$$

$$\mathbf{H}_{44} = \begin{bmatrix} \dfrac{\bar{k}_a L_b}{3} + E_b S_b & \dfrac{\bar{k}_a L_b}{6} \\ \dfrac{k_{a1} L_a}{6} & \dfrac{k_{a1} L_b}{3} + E_b S_b \end{bmatrix}$$

$$(\bar{k}_a = k_a \cos^2\theta + k \sin^2\theta) \tag{A12}$$

The Q-vectors in Eqs. (102-105) are given below

$$\mathbf{Q}_1^m = \left\{ -k_0 \quad (-1)^m k \quad -D_A \lambda_{Am}^2 \quad (-1)^m D_A \lambda_{Am}^2 \quad (-1)^m k \cos^3\theta \quad 0 \quad 0 \quad 0 \quad 0 \quad 0 \quad (-1)^{m+1} k \sin^3\theta \quad 0 \right\}^{\mathrm{T}} \tag{A 13}$$

$$\mathbf{Q}_2^m = \left\{ 0 \quad -k\cos\theta \quad 0 \quad 0 \quad -\bar{k} \quad (-1)^m k_1 \quad -D_B \lambda_{Bm}^2 \quad (-1)^m D_B \lambda_{Bm}^2 \quad 0 \quad -k_a \sin\theta \quad 0 \quad 0 \right\}^{\mathrm{T}} \tag{A 14}$$

$$\mathbf{Q}_3^m = \{0 \quad 0 \quad 0 \quad 0 \quad (-1)^{m+1}k_a\sin^3\theta \quad 0 \quad 0 \quad 0 \quad k_{a0} \quad (-1)^{m+1}k_a \quad (-1)^{m+1}k_a\cos^3\theta \quad 0\}^{\mathrm{T}} \tag{A 15}$$

and

$$\mathbf{Q}_4^m = \{0 \quad -k\sin\theta \quad 0 \quad 0 \quad 0 \quad 0 \quad 0 \quad 0 \quad k_a\cos\theta \quad \bar{k}_a \quad (-1)^{m+1}k_{a1}\}^{\mathrm{T}}. \tag{A 16}$$

The matrices and vectors used in Eqs. (109) and (110) are defined as

$$S_{rs,mm'}^{p(r)q(r)} = \mathbf{P}_m^{pq}\widetilde{\mathbf{H}}_r\mathbf{Q}_{sm'} \; , \qquad (p = A,\, B;\; q = w,\, u;\; r,\, s = 1,2,3,4) \tag{A17}$$

and

$$Z_{rs,mm'}^{p(r)q(r)} = \mathbf{Q}_{rm}^T\widetilde{\mathbf{H}}_r^T\Xi_q^p\widetilde{\mathbf{H}}_r\mathbf{Q}_{sm'} \tag{A18}$$

where

$$p(r) = \begin{cases} w\,, & \text{for } r = 1,2 \\ u\,, & \text{for } r = 3,4 \end{cases} \tag{A19}$$

$$q(r) = \begin{cases} A\,, & \text{for } r = 1,3 \\ B\,, & \text{for } r = 2,4 \end{cases} \tag{A20}$$

$$\mathbf{P}_m^{wq} = \begin{cases} \dfrac{2}{L_q}\left\{1/\lambda_{qm}{}^4 \quad (-1)^{m+1}/\lambda_{qm}{}^4 \quad -1/\lambda_{qm}{}^2 \quad (-1)^m/\lambda_{qm}{}^2\right\}, & m \neq 0 \\[2ex] \dfrac{2}{L_q}\left\{-1\; 1\; 0\; 0\right\}\,, & m = 0 \end{cases}, \tag{A21}$$

$$\mathbf{P}_m^{uq} = \begin{cases} \dfrac{2}{L_q}\left\{-1/\lambda_{qm}{}^2 \quad (-1)^m/\lambda_{qm}{}^2\right\}, & m \neq 0 \\[2ex] \dfrac{2}{L_q}\left\{-1\; 1\right\}\,, & m = 0 \end{cases}, \tag{A22}$$

$$\Xi_q^w = 2/L_q\int_0^{L_q}\zeta_{wq}(x)^T\zeta_{wq}(x)dx$$

$$= \begin{bmatrix} \dfrac{2L_q^{\,6}}{4725} & & & \\[2ex] \dfrac{127L_q^{\,6}}{302400} & \dfrac{2L_q^{\,6}}{4725} & \quad sym. & \\[2ex] -\dfrac{4L_q^{\,4}}{945} & -\dfrac{31L_q^{\,4}}{7560} & \dfrac{2L_q^{\,2}}{45} & \\[2ex] -\dfrac{31L_q^{\,4}}{7560} & -\dfrac{4L_q^{\,4}}{945} & \dfrac{7L_q^{\,2}}{180} & \dfrac{2L_q^{\,2}}{45} \end{bmatrix} \qquad (A23)$$

and

$$\Xi_q^u = 2/L_q \int_0^{L_q} \zeta_{uq}(x)^T \zeta_{uq}(x)\,dx$$

$$= \begin{bmatrix} \dfrac{2L_q^{\,2}}{45} & \dfrac{7L_q^{\,2}}{180} \\[2ex] \dfrac{7L_q^{\,2}}{180} & \dfrac{2L_q^{\,2}}{45} \end{bmatrix}. \qquad (A24)$$

Additionally, the following identities have been used in deriving Eq. (109)

$$S_{11,0m'}^{wA} + \lambda_{Am'}^4 S_{11,m'0}^{wA} \equiv 0, \qquad (A25)$$

$$S_{22,0m'}^{wB} + \lambda_{Bm'}^4 S_{22,m'0}^{wB} \equiv 0 \quad , \qquad (A26)$$

$$S_{33,0m'}^{uA} - \lambda_{Am'}^2 S_{33,m'0}^{uA} \equiv 0 \quad , \qquad (A27)$$

and

$$S_{44,0m'}^{uB} - \lambda_{Bm'}^2 S_{44,m'0}^{uB} \equiv 0 \ . \qquad (A28)$$

ACKNOWLEDGMENTS

The authors gratefully acknowledge the support of the NSF Grants (CMS-0528263, CMMI-0827233) under the supervision of Dr. Eduardo Misawa.

REFERENCES

[1] Leissa, A. W. *Vibration of Plates*; Acoustical Society of America, 1993.

[2] Doyle, J. F. *Wave Propagation in Structures*; Springer: Berlin, 1989.

[3] Ahmida, K. M.; Arruda, J. R. F. *Int. J. Solids Struct.* 2001, 38, 1669-1679.

[4] Igawa, H.; Komatru, K.; Yamaguchi, I.; Kasai, T. *J. Sound Vib.* 2004, 277, 1071-1081.

[5] Rao, C. K.; Mirza, S. *J. Sound Vib.* 1989, 130, 453-465.

[6] Maurizi, M. J.; Rossi R. E.; Reyes, J. A. *J. Sound Vib.* 1976, 48, 565-568.

[7] Abbas, B. A. H. *J. Sound Vib.* 1984, 97, 541-548.

[8] Goel, R. P. *J. Sound Vib.* 1976, 47, 9-14.

[9] Gutierres, R. H.; Laura, P. A. A. *J. Sound Vib.* 1996, 195, 353-358.

[10] Grossi, R. O.; Arenas, B. D. V. *J. Sound Vib.* 1996, 195, 507-511.

[11] Wang, J. T.-S.; Lin, C.-C. *J. Sound Vib.* 1996, 196, 285-293.

[12] Auciello, N. M. *J. Sound Vib.* 1996, 192, 905-911.

[13] Ruta, P. *J. Sound Vib.* 1999, 227, 449-467.

[14] Tolstov, G. P. *Fourier Series*; Prentice-Hall: Englewood Cliffs, NJ, 1965.

[15] Li, W. L. *J. Sound Vib.* 2000, 237, 709-725.

[16] Li, W. L. *J. Sound Vib.* 2001, 246, 751-756.

[17] Li, W. L. *J. Sound Vib.* 2002, 255, 185-194.

[18] Blevins, R. D. *Formulas for Natural Frequency and Mode Shape*; Van Nostrand Reinhold Company: New York, NY,1979.

[19] Bercin, A. N.; Langley, R. S. *Comp. Struct.* 1996, 59, 869-875.

[20] Langley, R. S. *J. Sound Vib.* 1990, 136, 439-452.

[21] Park, D. H.; Hong, S. Y.; Kil, F. G.; Jeon, J. J. *J. Sound Vib.* 2001, 244, 651-668.

[22] Keane, A. J.; Price, W. G. *J. Sound Vib.* 1991, 144, 185-196.

[23] Keane, A. J. *Proc. R. Soc. Lond. A* 1992, 436, 537-568.

[24] Beshara, M.; Keane, A. J. *J. Sound Vib.*1997, 203, 321-339.

[25] Shankar, K.; Keane, A. J. *J. Sound Vib.* 1995, 180, 867-890.

[26] Farag, N. H.; Pan, J. *J. Acoust. Soc. Am.* 1997, 102, 315-325.

[27] Keane, A. J. *Proc. R. Soc. Lond. A* 1992, 436, 537-568.

[28] Lee, H. P. *J. Sound Vib.* 1994, 171, 361-368

[29] Zheng, D. Y.; Cheung, Y. K.; Au, F. T. K.; Cheng, Y. S. *J. Sound Vib.* 1998, 212, 455-467.

[30] Zhu, X. Q.; Law, S. S. *J. Sound Vib.* 1999, 228, 377-396.

[31] Lee, H. P. *Appl. Acoust.* 1996, 47, 319-330.

[32] Cha, P. D. *J. Sound Vib.* 2005, 286, 921-939.

[33] Hamada, R. *J. Sound Vib.* 1981, 74, 221-233.

[34] Goel, R. P. *J. Sound Vib.* 1976, 47, 9-14.

[35] Hong, S.-W.; Kim, J.-W. *J. Sound Vib.* 1999, 227, 787-806.

[36] Chang, T.P.; Chang, F.I.; Liu, M.F. *J. Sound Vib.* 2001, 240, 769-778.

[37] Dowell, E.H. *J. Appl. Mech.* 1979, 46, 206-209.

[38] Gürgöze, M. *J. Sound Vib.* 1998, 217, 585-595.

[39] Posiadała, B. *J. Sound Vib.* 1997, 204, 359-369.

[40] Nicholson, J. W.; Bergman, L. A. *J. Eng. Mech.* 1986, 112, 1-13.

[41] Bergman, L. A.; McFarland, D. M. J. Vib., Acoust., *Stress* Rel. Des.1988, 110, 485-592.

[42] Abu-Hilal, M. *J. Sound Vib.* 2002, 267, 191-207.

[43] Kukla, S. *J. Sound Vib.* 1997, 205, 355-363.

[44] Foda M. A.; Abduljabbar, Z. *J. Sound Vib.* 1998, 210, 295-306.

[45] Leung, A. Y. T.; Zeng, S. P. *J. Sound Vib.* 1994, 177, 555-564.

[46] Henchi, K.; Fafard, M.; Dhatt, G.; Talbot, M. *J. Sound Vib.* 1997, 199 33-50.

[47] Wang, R.-T.; Lin, J.-S. *J. Sound Vib.* 1998, 212, 417-434.

[48] Yau, J. D.; Wu, Y. S.; Yang, Y. B. *J. Sound Vib.* 2001, 248, 9-30.

[49] Yang, Y. B.; Lin, C. L.; Yau, J. D.; Chang, D. W. *J. Sound Vib.* 2004, 269, 345-360.

[50] Henchi, K.; Fafard, M. *J. Sound Vib.* 1997, 199, 33-50.

[51] Dugush, Y. A.; Eisenberger, M. *J. Sound Vib.* 2002, 254, 911-926.

[52] Warburton, G. B. Proc. Inst. Mech. Eng. *A*, 1954, 168, 371-384.

[53] Leissa, A. W. *J. Sound Vib.* 1973, 31, 257-293.

[54] Dickinson, S. M.; Li, E. K. H. *J. Sound Vib.* 1982, 80, 292-297.

[55] Warburton, G. B. J. Earthquake Eng. Struct. *Dyn.* 1979, 7, 327-334.

[56] Warburton, G. B.; Edney, S. L. *J. Sound Vib.* 1984, 95, 537-552.

[57] Li, W. L.; Zhang, X. F.; Du, J. T.; Liu, Z. G. *J. Sound Vib.* 2009, 321, 254-269.

[58] Zhang, X. F.; Li, W. L. *J. Sound Vib.* 2009, 326, 221-234.

[59] Li, W. L. *J. Sound Vib.* 2004, 273, 619-635.

[60] Khov, H.; Li, W. L.; Gibson, R. F. *Compos. Struct.* 2009, 90, 474-481.

In: Fourier Transform Infrared Spectroscopy
Editor: Oliver J. Rees, pp. 79-99

ISBN: 978-1-61668-835-6
© 2010 Nova Science Publishers, Inc.

Chapter 3

FIBER OPTIC-COUPLED GRAZING ANGLE PROBE-FOURIER TRANSFORM REFLECTION ABSORPTION INFRARED SPECTROSCOPY FOR ANALYSIS OF ENERGETIC MATERIALS ON SURFACES

O. M. Primera-Pedrozo[*]*, Y. M. Soto-Feliciano,*
*L. C. Pacheco-Londoño, and S. P. Hernández-Rivera**
Center for Chemical Sensors Development
ALERT DHS Center of Excellence for Explosives
Department of Chemistry
University of Puerto Rico-Mayagüez, Mayagüez, PR 00681

ABSTRACT

This chapter focuses on the use of Fiber Optic-Coupled Grazing Angle Probe/Fourier Transform Reflection Absorption Infrared Spectroscopy (FOC-GAP/FT-RAIRS) as a new and promising technique for detecting and characterizing residues of neat energetic materials and their mixtures on stainless steel, glass and plastic surfaces. Smearing and TIJ techniques were used for transferring the target analytes to the substrates to be used as standards and samples. The sample transfer methods gave good sample distribution, reduced sample loss upon transfer and were easy to manipulate, giving good reproducible distributions. Samples with surface concentrations ranging from micrograms/cm^2 to nanograms/cm^2 of the explosives 2,6-dinitrotuelene (DNT), 2,4,6-trinitrotoluene (TNT), octogen (HMX), Tetryl, pentaerythritol tetranitrate (PETN), triacetone triperoxide (TATP), and PETN/2,4,6-TNT mixtures were deposited on glass, stainless steel or plastic. Methanol was used as a transfer solvent for the smearing sample preparation. Data were analyzed using chemometrics routines, specifically partial least squares (PLS). The methodology is remotely sensed *in situ* and can detect nanograms of the compounds. It is solvent free and requires neither sample preparation nor pre-treatment. The results show that detection limits as low as 10 ng/cm^2 can be detected using FOC-GAP/FT-RAIRS.

INTRODUCTION

In the last few years, researchers worldwide have focused their attention on the search for new technologies for the detection and characterization of chemical agents, energetic materials and narcotics in different environments and scenarios. In the case of in situ field detection of energetic compounds, the technique generally used is ion mobility spectrometry (IMS). The major advantages of IMS are its sensitivity in the picogram range, its continuous real time monitoring capability, reasonable price due to instrumental simplicity and ease of automation [1]. However, in general terms, IMS has a limited linear range and cannot be used for quantitative analysis. Also, analyte and background responses exhibit variations that occur with different reactive gas compositions and sample compositions [1]. Moreover, it is relatively easy to overload an IMS. Therefore, sample mass and size must be limited [2].

Optical spectroscopy, in countless platforms and applications, including absorption, emission, fluorescence, scattering and laser spectroscopy, is routinely used for the measurement of many different species at trace levels. However, optical spectroscopy has not been extensively applied in the energetic materials detection arena. This is due, in part, to physical constraints such as low vapor pressure, limited sample size, concealment, interferences and in part, to the spectroscopic characteristics of the compounds themselves [3].

Spectroscopic techniques have the potential to provide the best selectivity for explosives and offer an information-rich fingerprint that allows for near unambiguous identification. In the 1990's, direct detection by infrared absorption spectroscopy was not possible because of the limited sensitivity of this method. In addition, the test materials had to be placed physically within the spectrometer's sample compartment for measurement. The production of fiber-optic cables (FOCs) that transmit in the mid-IR range made possible the development of a range of spectroscopic probes for in situ analysis [4], [5]. FTIR spectroscopy can now be effectively used outside the confinement of the sample compartment, making it available for field work.

Sensors can be constructed in the MIR using any of the five basic sensing schemes: transmission, reflection, grazing angle (GA) reflection, attenuated total reflection (ATR), and a variant of the ATR effect known as the fiber evanescent wave [6]. Reflection-absorption infrared spectroscopy (RAIRS) operating at the grazing angle is an excellent optical absorption technique available for measuring low concentrations of chemical compounds adhered to reflective surfaces such as metals [7]. MIR spectroscopy operating at the ''grazing angle'' of incidence (approximately 80° from the surface normal) is considered one of the most sensitive optical absorption techniques available for measuring low concentrations of chemical compounds deposited on surfaces such as metals, glasses and plastics [7].

The establishment of RAIRS as an active area in the trace detection of chemicals is the result of a combination of high extinction coefficients in the MIR and the optical advantages of working at the grazing angle [8]. By combining a grazing angle head or probe (GAP) with a fiber optic cable that transmits in the MIR, the instrumentation becomes a platform for developing methodologies for real time, remotely sensed, in situ analysis [4]], [9].

GAP-FTIR operating in RAIRS mode with a sensing probe coupled to fiber optics has been used for the detection of active pharmaceutical ingredients (APIs) on metals [9] and on glass surfaces [10]. In these reports, combined teams from academia and the private sector demonstrated that the technique is an excellent alternative for the validation of cleanliness of metal walls of pharmaceutical reactors. Low limits of detection (LOD) ranging from 10 to 50 ng/cm2 of single API were achieved [9]. In addition, the methodology was applied to quantify APIs in mixtures and was demonstrated to not depend solely on the reflective properties of metallic surfaces, making detection on other surfaces such as glass and plastic surfaces viable [10], [11]. These characteristics of the methodology offer clear advantages over the traditional, time consuming, laborious and operator driven, swab-based HPLC method of analysis routinely used by the pharmaceutical industry in the validation of the cleanliness of batch reactors and other vessels and pipes.

In this chapter, we have reviewed the applications of Fourier transform reflection absorption infrared spectroscopy (FT-RAIRS) combined with a GA probe used outside the boundaries of sample compartments for the detection of organic peroxide- [12] and nitro-based energetic materials on stainless steel, plastics and glasses [13]-[15]. The capability of the technology for the characterization and detection of mixtures [16] is also discussed. The attractive features of this technique include portability, a simple and rugged design, high sensitivity and short analysis time. These features lead to a potential use of MIR-coupled FTIR for national security and defense applications such as airport screening and applications within the military, both for energetic materials detection and characterization, as well as in decontamination applications.

SAMPLES AND STANDARDS PREPARATION

During several years of working in samples and standards development for use in the spectroscopy-based detection of analytes on surfaces, several strategies and methodologies for transferring solid samples onto substrates were tested. Initial work consisted of transferring micropipettes (0.1 – 25 µL) with surface loadings of 10-100 $\mu g/cm^2$ [5] [18]. Then, aerosol air brush dispensing was tested in order to achieve a more uniform distribution and to avoid (or stimulate) crystallization of the analyte on the surface [10]. In another application, RDX samples were transferred to glass and silicon substrates using pneumatically assisted nebulization (PAN) with an electrospray needle [19]. This type of solid sample transfer was capable of biasing the α-RDX/β-RDX distribution and established a deposition method that offered control over the crystalline form of the RDX material being deposited. The methodology was validated using a fluorescent dye as target analyte transferred onto the test surfaces. The amount of dye transferred to the surfaces was recovered by rinsing and analyzed for total mass using fluorescence spectroscopy [19].

Two methods were developed for the preparation of homogeneous samples and standards of solids/traces deposited on surfaces. The applications of these specially prepared samples include use in experiments that require fine control of the distribution of analytes on surfaces. Another important application is in the establishment of reliable standards that may support an instrument response validation program. Smearing [10] [12]-[16] and thermal inkjet (TIJ) [17] sample deposition have been demonstrated to be good methods for transferring explosives on stainless steel (SS), plastic or glass surfaces. The sample smearing preparation generally follows the preparation of a standard solution in a solvent, in which the test chemical is highly soluble. Briefly, the analyte to be dispensed is first weighted in a high enough amount and then dissolved in the appropriate solvent so that a concentrated stock solution of the chemical can be prepared. Dilutions are performed in order to have the desired surface concentration after depositing 20 μL of the solution on the surface (an effective area of 46.3 cm^2, a 3 cm width and 15.4 cm length were used). Finally, 20 μL of the solution are smeared over the surfaces using a Teflon sheet that is inclined towards the right or left in a single pass operation as shown in Figure 1. The resulting deposit is air-dried at room temperature for 10-15 min to allow solvent evaporation, and then FT-RAIRS analysis is carried out. Since the methodology is based on gravimetric procedures, the samples used as standards may be considered as primary standards.

TIJ technology offers advantages when compared to smearing deposition. This method is not subject to human errors and gives more uniform coverage. Moreover, the surface loading concentration can be varied by changing the numbers of passes delivered to the sample, the dispensing frequency, and the applied energy and pen architecture. Also, the method includes precise delivery of a number of droplets with well characterized mass and concentration. Furthermore, only one solution need be used, avoiding dilutions that can increase the analytical errors caused by human intervention [17]. In the thermal inkjet method, a thin film resistor superheats less than 0.5% of the fluid in the chamber to form a gas bubble. This bubble rapidly expands (in less than ten microseconds) and forces a drop to be ejected through the exit orifice [20]. The samples were dispensed using an ImTech™ Imaging System, Corvallis, OR, model I-Jet 312S, (Figure 1) equipped with an HP™, Hewlett-Packard Caribe, LTD, Aguadilla, PR, model 51645a inkjet cartridge. Explosives solutions were placed into the inkjet cartridge and the backpressure was set to 3 inches of water using an external backpressure controller. The solutions were dispensed over SS plates using a zero dot spacing script (space between drops using HP inkjet) at a printing resolution of 600 dots per inch (dpi). Once the solvent had evaporated, the spectra of the samples were collected under the same conditions as the background plates (blanks). Thermal inkjet depositions were more uniform and were quantitative in gravimetric terms, but the technique is expensive and limited in terms of the properties of the solvent used (analyte solubility, solvent viscosity and surface tension).

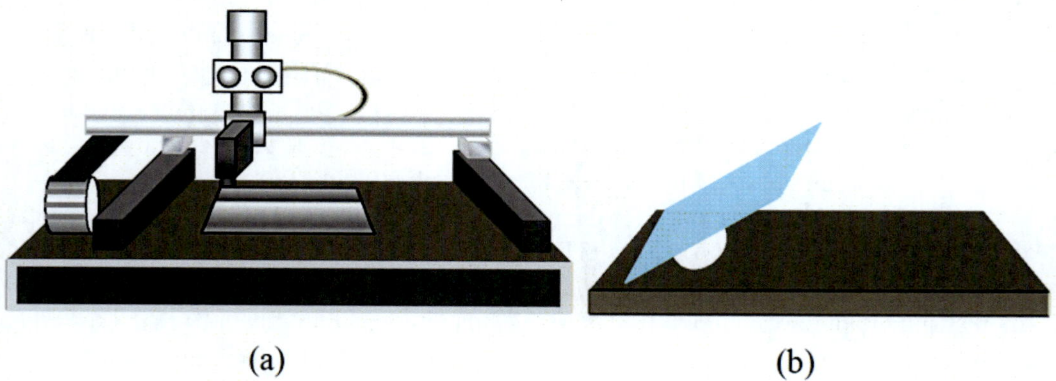

(a)   (b)

Figure 1. Methods transfer of solid sample onto a substrate: (a) Thermal Inkjet depositing; (b) Sample smearing.

INSTRUMENTAL SETUP AND SPECTROSCOPIC PARAMETERS

The setup for FOC-FTIR-RAIRS experiments has been described previously in contributions of our research group [13], [16]. A Remspec Corp. (Charlton, MA) grazing angle probe (GAP) accessory was connected to the external beam port of a Bruker VECTOR 22 FT-IR spectrometer by a round 19 optical chalcogenide fiber bundle of 18/1 configuration, 1.5 m long and 3 mm outer diameter. The setup for the experiments is schematically presented in Figure 2.

The As–Se–Te-based fiber optics cable used transmitted IR signals throughout the mid-infrared (MIR) region, with the exception of a strong H–Se absorption band centered at approximately 2200 cm^{-1} and a low wavenumber cutoff of about 900 cm^{-1}. The GAP head was carefully aligned using two off-axis parabolic gold coated mirrors that were used to collimate the IR beam exiting the interferometer and direct it to the sampling surface near the grazing angle (~80° from test surface normal). The reflected beam was then collected and returned to an MIR detector.

An external, liquid nitrogen-cooled, mercury–cadmium–telluride (MCT) detector was used to detect the reflected MIR light in the wavenumber range of 900–4000 cm^{-1}. The GAP head illuminated a large spot on the sample surface, forming an ellipse with dimensions of 1 inch by 6 inches. The IR beam at the sample had a Gaussian distribution of intensities with a center spot of maximum intensity of about 1/8 inch by 1 inch.

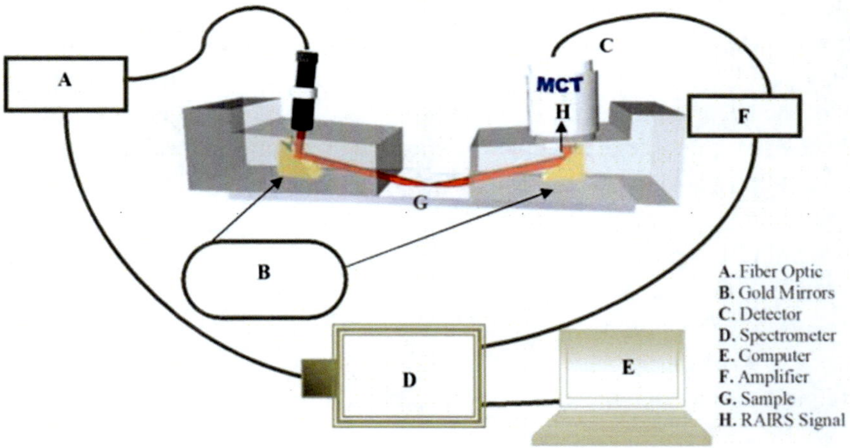

Figure 2. Experimental setup: Samples were deposited on the surfaces. Gold coated elliptical mirrors direct IR light to and from the samples. The grazing angle probe (GAP) is fiber-coupled to the FTIR spectrometer. The IR signal is detected by an MCT detector, amplified and sent to the interferometer. Data collection and analysis is controlled by a PC [13], [16].

The MIR intensity profile decayed from the middle towards the edges of the optical ellipse. The electric signal from the MCT detector was delivered to the interferometer using an amplifier.

FTIR experimental conditions are commonly co-addition of 50 scans, a 4 cm^{-1} resolution and a spectral range of 900–4000 wavenumbers (cm^{-1}). Since FT-RAIRS is a single beam technique, a background spectrum from clean test plates under the same instrumental conditions as the sample spectra had to be acquired prior to sample measurements. All spectra were recorded in absorbance mode to facilitate data processing. The clean test plates used as blanks for acquiring the background spectra were made of stainless steel [12]-[13], [15]-[17], glass, or plastics [14].

A minimum of 10 RAIRS spectra (Rs) were acquired for each loading concentration using the Bruker Optics OPUS™ software package for data collection. The calibration models were built with the QUANT2 package from Bruker Optics OPUS™ software using PLS1, a partial least squares (PLS) algorithm.

SOLVENT SELECTION FOR SURFACE DEPOSITING

The generation of vibrational chemical images allows determination of the distribution of explosives on the test surfaces. These images were constructed from data acquired in micro-RAIRS mode (IR mapping mode) in a 650–3600 cm^{-1} wavenumber range. They represent averages of 32 scans at a 4 cm^{-1} resolution within 1 mm x 1 mm grids separated by 100 µm. During this type of analysis, one band is selected for generation of the vibrational image. Peak areas are calculated and represented in a rainbow contour scale in which the x–y axes are spatial dimensions (µm). The spatial distributions were based on the variations or contrasts in pixel intensities.

In the case of nitro-based energetic materials, the asymmetric stretching vibration of the nitro group [v_{as} (NO_2)] was selected. Each pixel of the image contains spectral information of the explosive analyzed. The contrasts in the image are mainly due to differences in the intensities of the FTIR spectra at specific wavenumbers, particularly variations in the intensity of the absorption band that represents a specific chemical component of the sample. Figure 3 shows an example of 2,4,6-TNT deposited using different solvents [13].

The vibrational chemical images represented in Figure 3 are two-dimensional slices of the real three-dimensional distributions: two spatial coordinates in a plane, with one vertical axis representing the concentration at a given area in the plane of the substrate examined. Colors were assigned to different levels of surface loadings. Differences in color intensity can be clearly observed. The three spectra were used to generate the regions of highest surface loadings and are arbitrarily represented by red pixels (medium high concentrations) and pink pixels (highest surface loadings). Three spectra from the lowest surface concentrations are represented by light blue pixels (medium low concentrations) to dark blue pixels (lowest surface loadings) in the image field. The "x" and "z" axes are in microns and the second vertical axes on the right of each vibrational image are color coded and contain the surface concentration levels ($µg/cm^2$). Therefore, the pink-red pixels are enriched in explosive surface concentration. In contrast, the blue pixels indicate a diminished target compound surface loading. A color that tends to pink-red indicates a higher concentration of the energetic material, and one that tends to blue represents lower concentration of the compound.

In addition, by calculating the peak areas and standard deviations (SD) of the selected spectroscopic range and taking in account the values of kurtosis and skewness of the distributions of surface concentrations compared to the normal (Gaussian) distribution, it is possible to establish the relative homogeneities of the distributions.

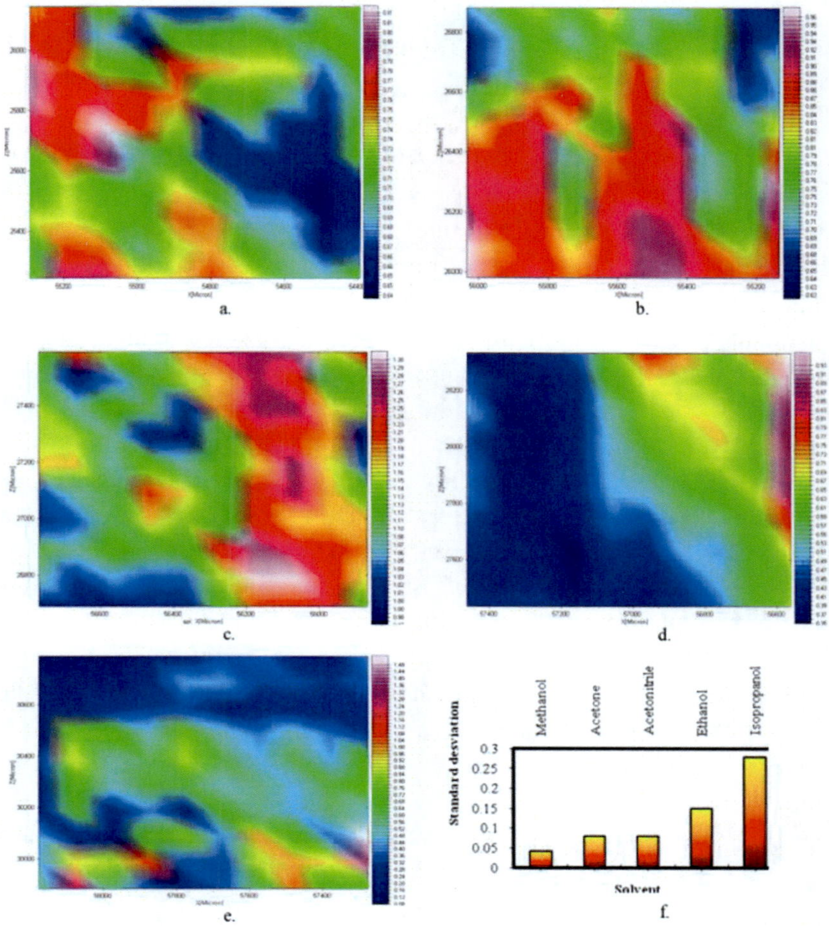

Figure 3. Infrared spectroscopic images of ~ 10 µg/ cm² of TNT deposited on a SS plate using different solvents: (a) methanol; (b) acetone; (c) acetonitrile; (d) ethanol; (e) isopropanol. (f) Plot of standard deviation of the peak areas in the images for each solvent [13].

Based on the low SD for methanol (Figure 3f), this solvent was selected as the best solvent for distributing 2,4,6-TNT on metallic surfaces [13]. This was the basis used for selecting methanol as the optimum solvent for the FOC-RAIRS experiments. A likely explanation for this finding is based on the high solubility of explosives in methanol and the high vapor pressure and evaporation rate at room temperature of methanol.

Volatile solvents in which nitroexplosives are soluble (such as acetone, acetonitrile and methanol) resulted in more uniformly distributed surfaces when they were deposited on the stainless steel surfaces. However, during the solvent evaporation stage, parts of the material tended to aggregate with the escaping solvent, forming island-type structures and generating surface concentration gradients (hills and valleys). This phenomenon was accentuated for solvents with low vapor pressure and evaporation rates, which gave way to the formation of islands due to crystallization of the analyte on certain parts of the substrate.

FOC-RAIRS OF NITRO AND NON-NITRO ENERGETIC MATERIALS ON METALLIC SURFACES

One clear disadvantage of FOC-GAP-RAIRS, as shown in Figure 2, is that the method is an open path experiment in which the IR beam passes through the atmosphere before and after interacting with the sample. As a result, some absorption bands from water vapor (3800–3400 and 1600 cm^{-1}) and CO_2 (~2350 cm^{-1}) are observed. The solution to this problem is the collection of a background single channel spectrum before each sample run. Another possible solution is the mathematical elimination of these bands using algorithms that are incorporated in standard FTIR data acquisition and analysis software packages, such as the Bruker Optics OPUS™ spectroscopic suite. However, the nitro bands in nitrogen-based high explosives (Figure 4) are very intense, so that these atmospheric bands are sometimes negligible. NO_2 asymmetric (~ 1600 cm^{-1}) and symmetric (~ 1250-1375 cm^{-1}) bands act as vibrational signatures for the discrimination of nitroaromatic HE (TNT, DNT, Tetryl), nitroaliphatic HE (PETN, HMX, RDX), nitramines (HMX, Tetryl) and nitrate esters (PETN).

As shown in Figure 4, the region of 1200 cm^{-1} to 1600 cm^{-1} is rich in spectroscopic information for nitrogen-based energetic materials, specifically in the region of the nitro symmetric stretch vibration (ν_{sym} ~ mid fingerprint region) and the asymmetric nitro stretch vibration (at the end of the fingerprint region). In PETN, the symmetric stretch band appears in the 1250–1320 cm^{-1} region. For energetic compounds such as TNT, DNT and HMX, ν_{sym} appears in the wavenumber range: 1320–1360 cm^{-1} [21].

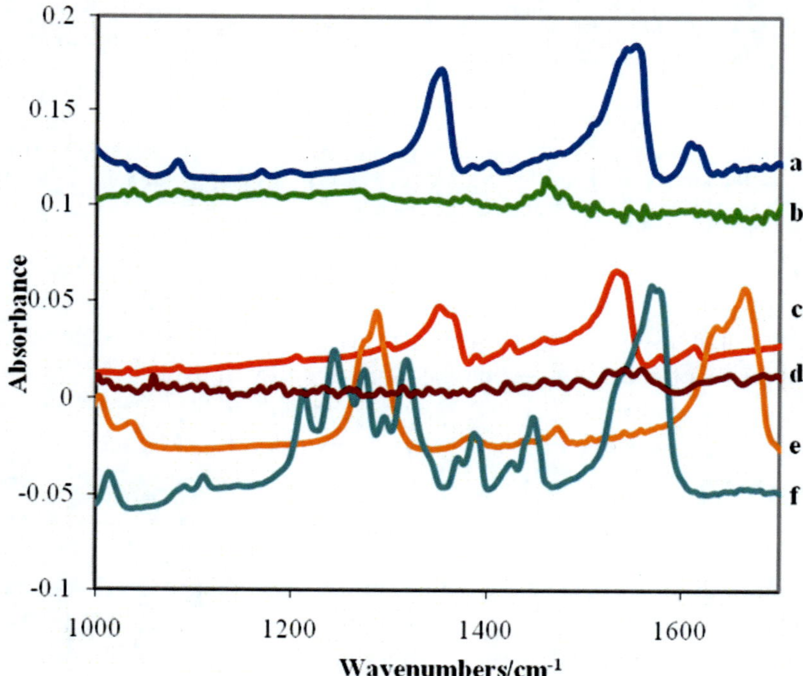

Figure 4. RAIRS spectra of nitro-based energetic materials on SS at different loading concentrations: (a) TNT, 5.0 µg/cm^2; (b) Tetryl, 8.1 µg/cm^2 (absorbance x3); (c) 2,6-DNT, 10.0 µg/cm^2; (d) Blank SS (absorbance x100) (e) PETN, 8.5 µg/cm^2; (f) HMX, 7.2 µg/cm^2 [13].

The difference in wavenumber location can be explained in terms of the fact that the NO_2 groups in PETN are attached to oxygen atoms. However, in TNT, the nitro groups are directly attached to carbon atoms in the aromatic moiety. The high electronegativity of the oxygen atom in PETN attracts electron density from the nitro group, leading to a decrease of the oscillator strength and a shift to lower frequencies. This effect is smaller or not present at all in the aromatic ring of TNT [16].

The capability of discrimination between neat energetic compounds and their mixtures has also been demonstrated using this technology. Detection and discrimination between the mix and its constituents of the well known formulation of TNT and PETN (Pentolite) was used as a proof of concept [16].

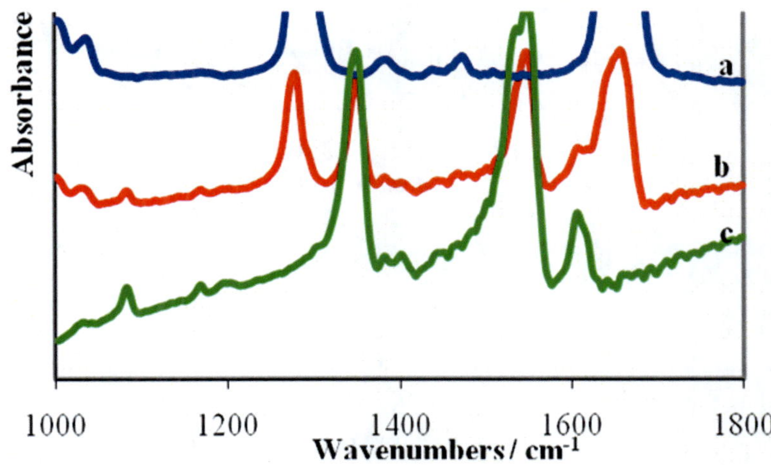

Figure 5. Grazing angle FTIR spectra for: (a) PETN; (b) PETN-TNT mixture; (c) TNT on SS [16].

The experiments demonstrated that both explosives have signature bands that can be distinguished from the rest of the components present in the sample in the case in which relatively low surface concentrations of the samples were present. As explained before, the nitro stretching vibrations of both explosives appear at different wavenumbers. So, it can be clearly observed that the spectra of the Pentolite mixtures contain features of both energetic material, and no significant overlapping of bands is observed in the spectra (Figure 5).

Experiments on non-nitrogen based energetic materials have also been reported [22]. Detection of organic peroxide compounds such as TATP, at trace levels as residues on surfaces using FOC-GAP-RAIRS (or any other methodology), represents a real challenge due to its low surface affinity and high sublimation rate even at room temperature. There are currently few methods for the *in situ* detection of this type of molecule, and despite its widespread misuse, only a small number of scientific studies have been published on the detection of this compound [22] - [24]. Despite the fact that TATP has a vapor pressure of 6.95-7.87 Pa at ambient conditions [25] - [26], the most important characteristic bands are observed, even at small loading concentrations.

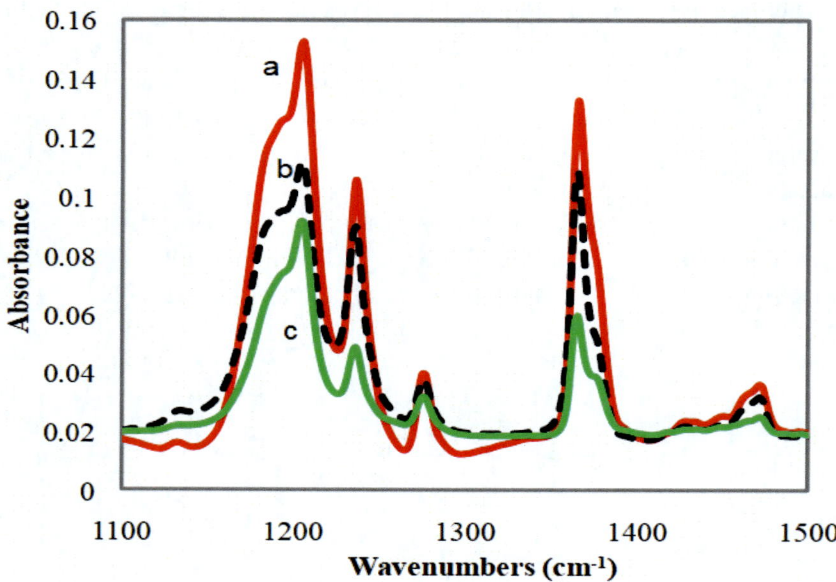

Figure 6. Grazing angle FT-IR spectra of TATP at different loading concentrations on SS: (a) 100 µg/cm^2; (b) 40 µg/cm^2; (c) 10 µg/cm^2.

Figure 6 illustrates that several TATP bands can be used for its accurate detection. The presence of prominent bands of TATP in the fingerprint region is clearly illustrated. The band at 1205 cm^{-1} belongs to the C-O stretch, 1365 cm^{-1} is a deformation of the CH$_3$ group and the band at 1471 cm^{-1} is an asymmetric deformation of the CH$_3$ group [21]. The amount of explosive on the surface of stainless steel and the residence time on the substrate can be accurately correlated with the IR intensity (band area) of its signature vibrational bands. At very low surface loading concentrations, the explosive rapidly goes into the vapor phase, limiting the precision of the results.

FOC-GAP-RAIRS SPECTRA OF 2,4,6-TNT DEPOSITED ON NON TRADITIONAL SURFACES

RAIRS signals depend on the type of substrate from which the radiation will be reflected. Commonly, substrates are classified as metallic [7], semiconductor, and dielectric [8]. When dielectric substrates are used, the RAIRS spectra look different when compared with spectra obtained on metallic surfaces [27]. An illustration of this effect is observed on a polyethylene (plastic sheet) substrate, as shown in Figure 7. When metallic substrates are used for RAIRS analysis, the grazing angle probe is much closer to the substrates due the lack of absorption by the metal surface. However, plastics strongly absorb in the mid-infrared. For this reason, the substrate to focal plane distance was optimized to 5 cm in order to minimize absorption by the substrate and the low intensity of the band located at 1466 cm^{-1}.

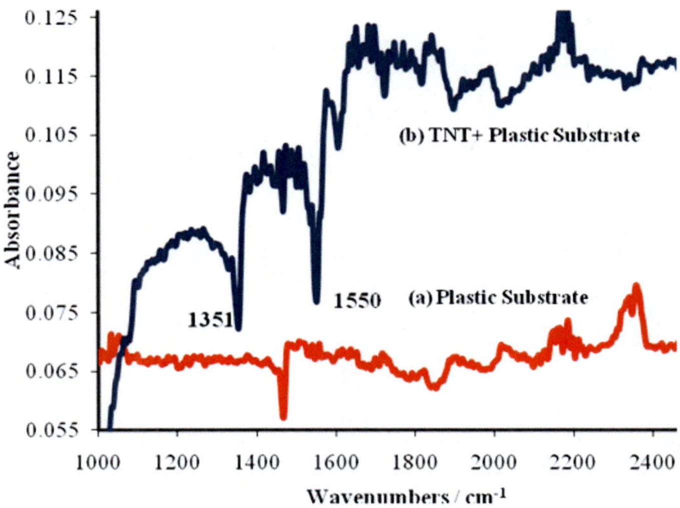

Figure 7. RAIRS spectra of: (a) plastic substrate; (b) TNT deposited on the plastic substrate [14].

The NO_2 symmetric stretch vibration is clearly observed in the spectrum (1320 – 1360 cm^{-1}). As expected for dielectric substrates, negative features are observed [8]. This is due to the fact that negative or positive bands depend on some properties of the substrates, such as the refractive index or the polarization of light, and the incidence angle [21]. The RAIRS spectrum on glass has similar features compared to the plastic surface (Figure 8). In both cases, inverted bands are observed. However, on glass there is the presence of a strong band at about 1267 cm^{-1} attributed to Si-O modes. This Si-O band intensity depends on the loading concentration on the substrate. The intensities in the non-metallic substrates present low absorbance in comparison with stainless steel [10]. However, the symmetric and asymmetric bands of the nitro vibration are very distinguished from the high absorption band of the glass substrate.

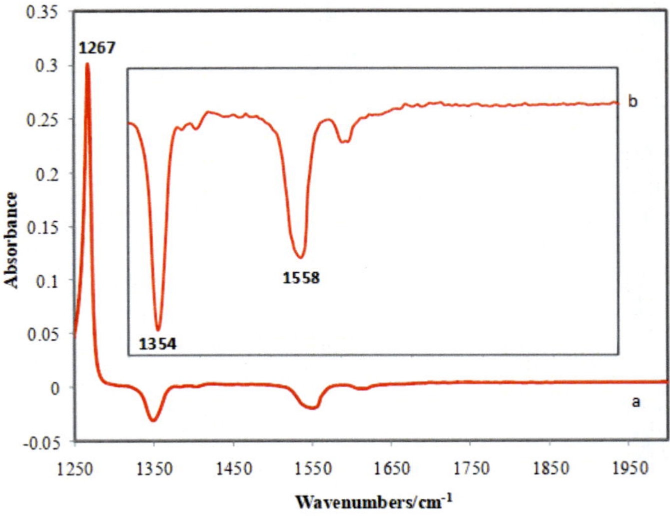

Figure 8. RAIRS spectra of 2,4,6-TNT deposited on a glass substrate: (a) RAIRS spectrum showing the SiO signal; (b) zoom of nitro group signatures.

CHEMOMETRICS MODELS AND DISCRIMINANT ANALYSIS

Calibrations are typically constructed using chemometrics methods of data analysis such as the partial least squares (PLS) regression algorithm. The PLS algorithm is commonly incorporated in spectroscopic software such as OPUS™, Pirouette™ and Matlab Toolboxes™, among others. PLS analysis was used to correlate the loading concentrations on the tested surfaces with the RAIRS spectra [10]. The advantages of using chemometrics for the quantification of organic compounds on glass, aluminum and stainless steel [12]-[17] surfaces have been discussed in the literature [11]. In addition, PLS is a multivariate method that uses spectral features over a wide spectroscopic range. It is a spectral decomposition method that is intimately related to principal component analysis (PCA). In PCA, the spectral matrix is first decomposed into a set of eigenvectors and scores, and then a regression is performed against the concentrations as a separate step. However, PLS uses the concentration information during the decomposition process [28].

In the case of OPUS™ (Bruker Optics), Quant2 software is used to find the best correlation function between the spectral information and the loading concentrations. Quant2 uses a partial least squares-1 (PLS-1) fit method. Calibrations are performed using PLS-1 in which only one component can be analyzed separately, instead of simultaneously analyzing multiple components, as in the PLS-2 routine of chemometrics. Then, cross validations are performed and the root mean square errors of the cross validations (RMSECV) and the root mean square errors of estimations (RMSEE) are used as criteria to evaluate the quality of the correlations obtained. In the standard "leave-out-one" cross validations, each spectrum is omitted from the training set and then tested against the model built with the remaining spectra. As illustrated in Tables 1 and 2, some explosives require spectroscopic preprocessing (except mean centering) and more PLS evaluations in order to obtain a good model. In the case of the first derivative, the Savitzky-Golay algorithm is actually used to obtain the derivative. The number of smoothing points used can be adjusted to suppress the effect of noise [29]. Other details of the advantages of using a chemometrics model such as PLS to correlate the loading concentration with the RAIRS spectra have been discussed in the literature [10].

$$RMSECV = \sqrt{\frac{\sum_{i+1}^{I}(c_i - c_p)^2}{I}}$$

c_i: true concentration
c_p: predicted concentration
I: number of samples

$$\tag{1}$$

$$RMSEE = \sqrt{\frac{1}{M - R - 1}SSE}$$

SEE: sum of squared errors
Res: difference between the true and fitted value
M: number of samples

$$\tag{2}$$

$$SEE = \sum_i [Res_i]^2$$

R: number of PLS (Rank)

$$\tag{3}$$

Table 1. Quant2 calibration parameters for the chemometrics models for the studied compounds on SS [13]

	2,4,6-TNT[a]	PETN[b]	2,6-DNT[c]	HMX[d]	Tetryl[e]	TATP
Loading concentration ($\mu g/cm^2$)	0.005-1.73	0.3-6.5	0.15-5.00	0.6-7.2	5.4-13.5	8.0-35.0
Spectral range (cm^{-1})	1028-1713	1878-1088	1057-1979	1632-968	3024-2810	1113-1498
Preprocessing	No	First derivative	First derivative	No	No	No
Rank (PLS)	3	10	9	8	7	3
RMSECV ($\mu g/cm^2$)	0.258	0.498	0.398	0.42	0.78	3.69
RMSEE	0.18	0.37	0.19	0.27	0.58	3.56
R^2	0.996	0.914	0.946	0.97	0.934	0.868
Number of samples	84	72	22	60	48	79

Table 2. Quant2 calibration parameters for the chemometrics models for 2,4,6-TNT on non-metallic surfaces [14]

Surface	Loading concentration ($\mu g/cm^2$)	Spectral range/ cm^{-1}	Preprocessing	Rank (PLS)	RMSECV ($\mu g/cm^2$)	RMSEE	R^2	Number of samples
Plastic[a]	0.08-5.0	1028-1713	No	4	0.323	0.308	0.998	105
Glass	0.3-5.0	1306-1888	No	5	0.364	0.342	0.967	68

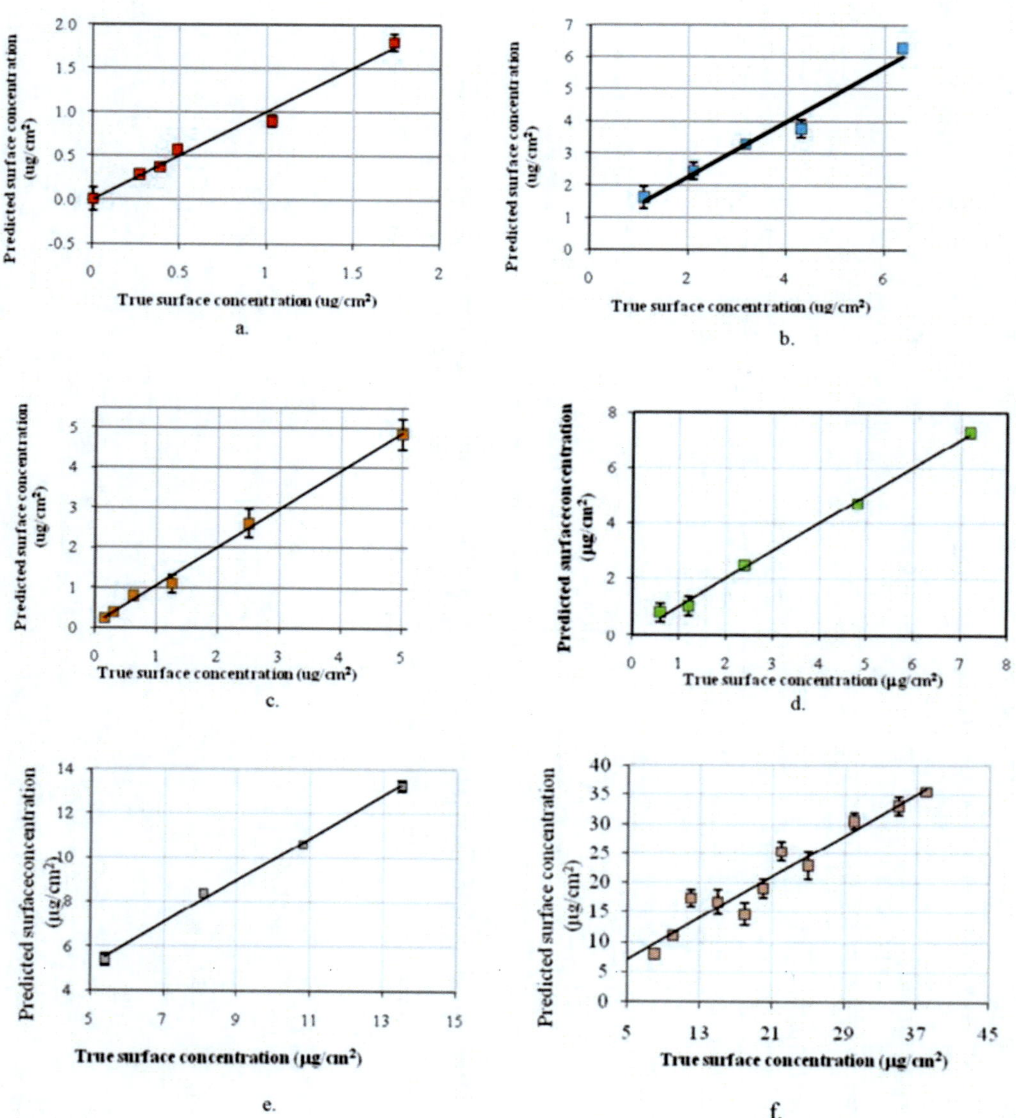

Figure 9. Predicted versus surface loading concentrations on SS for the studied explosives from a leave-one-out cross validation using Quant2: (a) 2,4,6-TNT; (b) PETN; (c) 2,6-DNT (d); HMX (e) Tetryl; [13] (f) TATP.

In the cross validations shown in Figures 9 and 10, true loading concentrations were plotted against predicted loading concentrations. Figure 9 contains the results obtained for a stainless steel surface for (a) 2,4,6-TNT, (b) PETN, (c) 2,6-DNT (d), HMX (e) Tetryl, [13] and (f) TATP.

Figure 10 contains the results for 2,4,6-TNT on plastic (a) and glass (b). The results of ''leave-one-out'' cross validation for each of the nitroexplosives studied, as well as the model parameters obtained are shown in Table 1. Low RMSECV values in the cross validations indicate the high-quality of the chemometrics models obtained that were used for the predicting loading concentrations of unknown explosives samples on metallic and non-metallic surfaces.

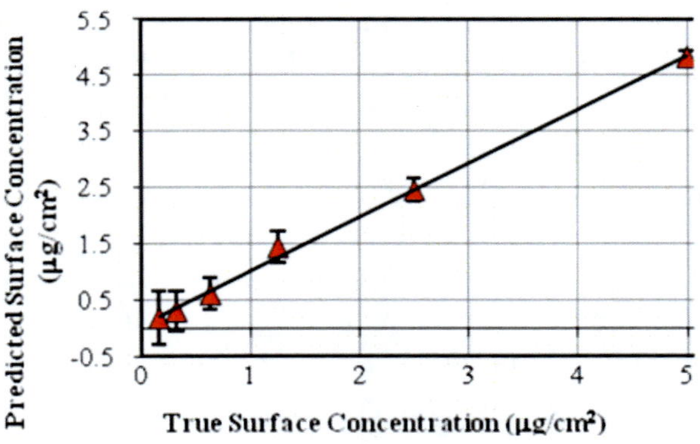

a.

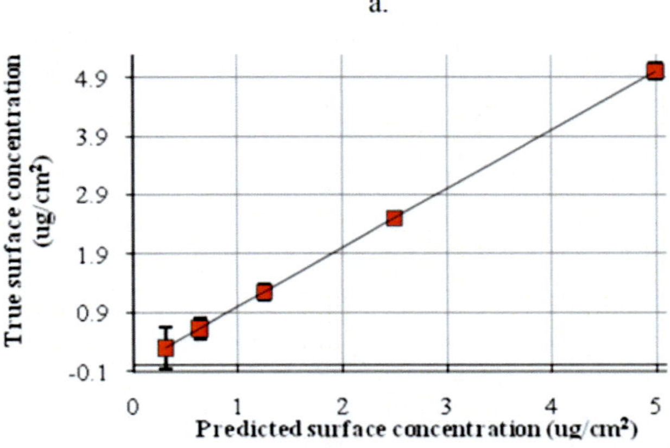

b.

Figure 10. Predicted versus surface loading concentrations of 2,4,6-TNT from a leave-one-out cross validation using Quant2: (a) plastic substrate; (b) glass substrate [14].

These values were in the low $\mu g/cm^2$ level to ng/cm^2 for nitroexplosives, but were slightly higher for Tetryl (Figure 9-e). Cross validations of TATP were significantly higher (3.69 $\mu g/cm^2$, Figure 9-f) for the reasons already discussed. The spectral range of the chemometrics models built cover the fingerprint region where the nitro symmetric stretch vibration is located for nitroexplosives. In the case of the aromatic nitramine Tetryl, the region used was 3024–2810 cm^{-1}, which corresponds to the aromatic CH region. When the cross validations for each explosive exhibited low ranks (ranging from 3 to 10), this indicated that satisfactory models were obtained.

However, not all of the spectroscopic regions give good information for the construction of calibration models. This was the case for TATP, where a PLS1 model was built from all of the 79 spectra listed for TATP using the spectral region from 1066-1506 cm^{-1}. With no spectral preprocessing, it proved impossible to develop a model that met a reasonable standard (the maximum obtainable value for R^2 was about 0.75). However, the R^2 value improved to 0.868 when the model parameters were optimized in the spectral region of 1498–1113 cm^{-1} without any spectral data preprocessing. For substrates with high absorption, the spectroscopy range was excluded for the calibration studies. Also, when non-reflective substrates were used, the spectroscopic range for the calibration models must not contain information about the substrate bands present, as shown in Figures 7 and 8 for plastic and glass surfaces, respectively (Table 2).

The rapid sublimation of TATP under ambient conditions presented a significant challenge in applying the fiber optic grazing angle method to surface detection of the compound. Even under controlled laboratory conditions, where spectroscopic measurements were carried out on freshly prepared samples, it proved impossible to develop PLS1 calibrations that met a high standard of robustness or usability. The rapid change in the surface concentration of TATP made it impractical to collect more than one spectrum from each sample, and this further limited the possibilities of building a statistically useful data set. While some of this scattering may be attributable to variations in the amount of TATP deposited on each coupon, sublimation of TATP during the experiment is likely another contributing factor. Given these limitations, the quality of the calibration that was developed is surprisingly good, and it is clearly quite possible to detect microgram quantities of TATP on metal surfaces using grazing angle FTIR methods. This is in agreement with previous results for a range of organic compounds on metal [9] and glass surfaces. Attempts to build a separate calibration for loadings of 40 $\mu g/cm^2$ and above were unsuccessful. This may be attributable to the change in the nature of the surface coating at high loadings from a thin film capable of generating a double-pass transmission spectrum to a bulk material generating diffuse surface reflectance independent of the thickness.

The limits of Detection (LOD) are considered to be the smallest concentrations that can be measured. According to IUPAC definitions, the limits of detection (LOD) are calculated according to Eq. 4 [30], [31]:

$$RSD = \frac{SDI}{SLOPE} x3 \qquad (4)$$

As the concentration of the sample decreases, the precision, expressed as the relative standard deviation (RSD), degrades. Values of RSD of 33% are considered as the limit of detection (k = 3 in IUPAC definition or three detection limits) [31]. Values of RSD were calculated for each PLS-1 model according to eq. 1, in which the standard deviation of the intercept (SDI) is divided by the slope multiplied by 3. According to Table 3, these values are markedly dependent on the type explosive on the surface. Properties such vapor pressure, physical adsorption, sublimation rate and surface-adsorbate thermodynamics influence the detection limit. Explosives such as TATP with high vapor pressure (5.25×10^{-2} Torr at 25 °C) give high LOD values when they are detected on surfaces, while nitroexplosives stay for a much longer time on surfaces.

Discriminant Analysis (DA) is another powerful statistical tool for data analysis of explosives on surfaces. DA is based on the classification of a set of observations into predefined classes. Discrimination is a form of classification. In spectral analysis, DA can be coupled to a statistical analysis routine that allows compression of the information contained in the multitude of spectral data points into fewer variables [29], [32]-[35]. The most popular of these methods is principal components analysis (PCA). The spectroscopic data can be fed into PCA and then the principal components (PC) scores (Figure 11) can be used as the input variables for linear discriminant analysis (LDA).

Figure 12 shows how the discriminant model can be used to analyze neat TNT, neat PETN, TNT-PETN mixtures (Pentolite) and the important case of when the energetic materials or their mixtures are not present on the substrate. As can be seen in Figure 11, the discriminant model improved when the numbers of scores increased. When the usual LDA method was applied using 20 principal components (PCs) as inputs and applying forward selection criterion, 9 PCs (scores) were selected: PC1, PC3, PC2, PC8, PC4, PC6, PC7, PC5 and PC14. It is clearly seen that PC1, PC3 and PC2 are adequate enough for a good classification (95.04%).

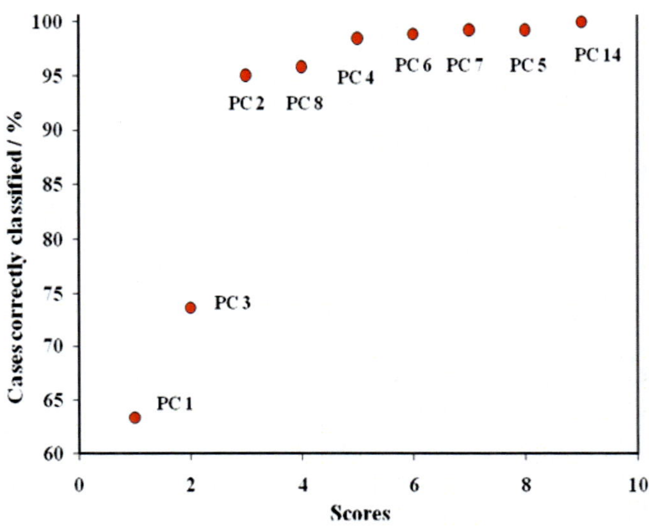

Figure 11. Improvement of the classification model as the number of scores (PC) increases [16].

Table 3. Limit of detection (LOD) of the studied nitro compounds and for TATP* with PLS1 [13], [14]

Explosive	Slope	Standard deviation of the intercept	LOD (ng/cm^2)*	Significance P value	Surface
2,6-DNT[a]	1.05	0.075	220	5.67×10^{-6}	SS
2,4,6-TNT[b]	1.00	0.05	160	7.5×10^{-5}	SS
PETN[c]	0.86	0.31	220	2.0×10^{-3}	SS
HMX[d]	0.99	0.13	400	7.61×10^{-5}	SS
Tetryl[e]	1.05	0.39	160	1.4×10^{-3}	SS
TATP[f]	1.04	2.22	800	2.1×10^{-6}	SS
2,4,6-TNT[g]	0.96	0.045	300	9.0×10^{-8}	Plastic
2,4,6-TNT[h]	0.96	0.08	600	8.1×10^{-5}	Glass

Three functions were derived. They were labeled as function one (F1), function two (F2), and function three (F3). For the TNT/PETN mixtures, the first function contains nearly all of the statistically relevant information, as they contributed to 93.4% of the discrimination capability. The p-value of 0.0001 of the eigenvalues is an indication that discriminant functions are highly significant. The canonical correlation coefficients of the functions give the capacities for determining group differences [16]. For the mixture, function one (F1) and function two (F2) have excellent capacities for determining group differences: 96% and 92%, respectively, of the ability to predict new samples. This can be better illustrated when F1 is plotted against F2, as depicted in Figure 12.

These values represent the percent of the squares of the canonical correlations and indicate the effectiveness of the two functions in the discrimination capability of the samples. When LDA was applied using the total number of principal components (9) as inputs, it can be observed that the two canonical variates (CV) or canonical roots separated the groups in a very clear fashion.

The TNT/PETN mixture (Pentolite) is well separated from the others. The best discrimination model was selected based on statistical significance and the percentage of cases correctly classified. The two discriminating functions with p-values less than 0.05 were statistically significant at the 95% confidence level. The percent of correctly classified cases was 100% [16].

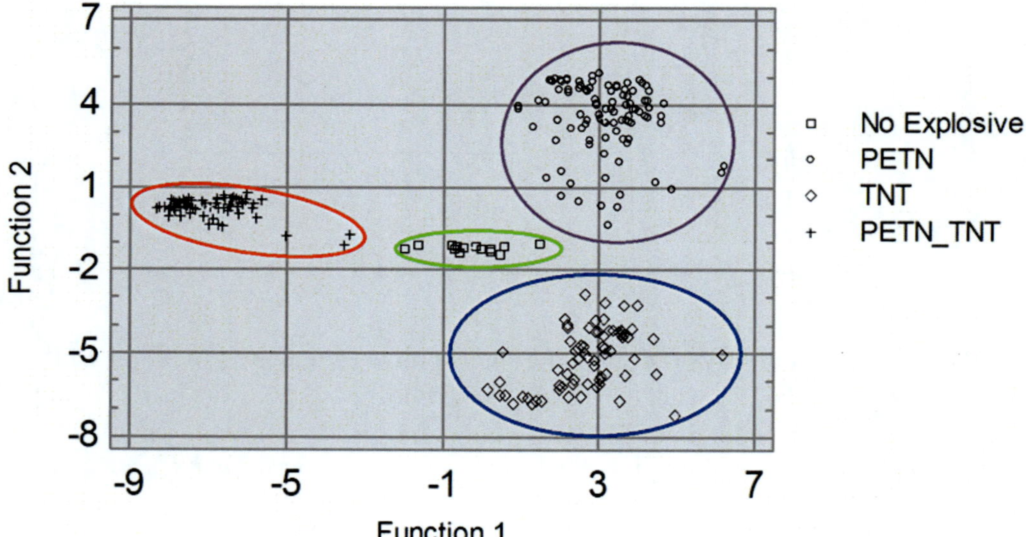

Figure 12. Discriminant functions plot of Function 1 and Function 2 using linear discriminant analysis (LDA) with 9 scores as inputs [16].

CONCLUSION

Fiber optic-coupled – grazing angle probe – FTIR spectroscopy is a recently developed technique that resulted from the combination of high throughput optical fibers in the mid-infrared region coupled to a compact grazing angle probe. The technique senses molecules deposited on surfaces and absorption-reflection phenomenon that takes place near the grazing angle (> 80° from the surface normal). FOC-GAP-FTIR has been shown to provide a useful framework for methods development for the *in situ* detection of traces of energetic compounds and mixtures of energetic materials on metallic, glass and plastic surfaces.

Very low limits of detection (160–800 ng/cm^2) were found for nitroexplosives, both nitroaromatic and nitroaliphatic. The technique is applicable for the trace detection of analytes on other types of surfaces and does not rely on surface reflectivity.

Two methods of dispensing analytes to prepare samples and standards in which a solid is deposited on a substrate with near uniform distribution were discussed. Sample smearing is inexpensive, easy to implement and provides sample loadings with relatively good distributions. When using a volatile solvent, such as methanol, acetone or diethyl ether, the formation of layers is favored.

Low volatility solvents tend to form islands because crystallization is faster than solvent evaporation. Thermal inkjet depositions are more uniform and quantitative in gravimetric terms, but the technique is expensive and limited in terms of the properties of solvent used (analyte solubility and solvent viscosity and surface tension).

Detection of target molecules using this technique was found to be limited by the residence time of the substance on the surface, which in turn depends on physical properties of the adsorbate and interaction forces with the surface. At low loading concentrations, high vapor pressure energetic materials quickly escape to the vapor phase by sublimation, limiting the limit of detection achievable.

This is the case of the peroxide-based homemade energetic compounds such as triacetone triperoxide (TATP), which has a very high sublimation rate at ambient temperatures and pressures. TATP can be detected fast (less than 10 s), but limits of quantification and low limits of detection are hampered by the properties of these unstable molecules.

ACKNOWLEDGMENTS

Parts of the work presented in this contribution were supported by the U.S. Department of Defense, University Research Initiative Multidisciplinary University Research Initiative (URI)-MURI Program, under grant number DAAD19-02-1-0257. The authors also acknowledge contributions from Mr. Aaron LaPointe from Night Vision and Electronic Sensors Directorate, Fort Belvoir, VA, Department of Defense, Dr. Jennifer Becker MURI Program Manager, Army Research Office, DoD and Dr. Stephen J. Lee Chief Scientist, Science and Technology, Office of the Director, Army Research Office/Army Research Laboratory, DoD.

Support from the U.S. Department of Homeland Security under Award Number 2008-ST-061-ED0001 is also acknowledged. However, the views and conclusions contained in this document are those of the authors and should not be interpreted as necessarily representing the official policies, either expressed or implied, of the U.S. Department of Homeland Security.

Thanks are due to Dr. Luis. F De La Torre-Quintana for his help with the statistical analysis and coupling with chemometrics routines and for his support and collaboration in the design and preparation of the figures.

REFERENCES

[1]   Salleras, L.; Donguez, A.; Mata, E.; Taberrer, J. L.; Moro, I.; Salva, P. *Public Health Rep.* 1995, 110, 338-342.

[2]   Brambilla, G.; Loizzo, A.; Fontana, L.; Strozzi, M.; Guarino; A.; Soprano, V. *J Am Med Assoc.* 1997, 278, 635.

[3]   Steinfeld, J. I.; Wormhoudt, J. *Annual Review of Physical Chemistry.* 1998, 49, 203–232.

[4]   Melling, P. J.; Shelley, P. "Spectroscopic Accessory for Examining Films and Coatings on Solid Surfaces" U.S Patent 6,3,10,348, United States Patent and Trademark Office, Washington, DC.

[5]   Mehta, N. K.; Goenaga-Polo, J. E.; Hernández-Rivera, S. P.; Hernández, D.; Thomson, M. A; Melling, P. *J. Bio Pharm.* 2002, 15, 36-42.

[6]   Mizaikoff, B. Sensory systems based on mid-infrared transparent fibers. In: J. M. Chalmers and P. R. Griffiths, Eds. Handbook of vibrational spectroscopy. Chichester, UK: Wiley and Sons; 2002, Vol. 2, pp. 1560–1573.

[7]   Griffiths, P. R.; De Haseth, J. A., Fourier- Transform Infrared Spectrometry. John Wiley, New York: 1986, pp. 194.

[8]   Umemura, J. Reflection–absorption spectroscopy of thin films on metallic substrates. In J. M. Chalmers and P. R. Griffiths, Eds., Handbook of vibrational spectroscopy. Chichester, UK: Wiley and Sons; 2002, Vol. 2, pp. 982–998.

[9]   Mehta, N. K.; Goenaga-Polo, J. E.; Hernández-Rivera, S. P.; Hernández, D.; Thomson, M. A; Melling, P. J., *Spectroscopy*, April, 2003.

[10]   Hamilton, M. L.; Perston, B. B.; Harland, P. W.; Williamson, B. E.; Thomson, M. A.; Melling, P. J. *Organic Process Research and Development.* 2005, 9, 337–343.

[11]   Perston, B. B.; Hamilton, M. L.; Williamson, B. E.; Harland, P. W.; Thomson, M. A.; Melling, P. J. *Analytical Chemistry.* 2007, 79, 1231–1236.

[12]   Primera-Pedrozo, O. M.; Pacheco Londono, L. C.; De la Torre-Quintana, L. F.; Hernandez-Rivera, S. P.; Chamberlain, R. T.; Lareau, R. T. *Proceedings of SPIE.* 2004, 5403, 237–245.

[13]   Primera-Pedrozo, O. M.; Soto-Feliciano, Y.; Pacheco-Londoño, L. C.; Hernandez-Rivera, S. P. *Sensing and Imaging: An International Journal.* 2009, 10, 1–13.

[14]   Primera-Pedrozo, O. M.; Rodríguez, N.; Pacheco-Londoño, L. C.; Hernández-Rivera, S. P. *Proceedings of SPIE.* 2007, 6542, 65423J.

[15]   Soto-Feliciano, Y.; Primera-Pedrozo, O. M.; Pacheco-Londoño, L. C.; Hernandez-Rivera, S. P. *Proceedings of SPIE.* 2006, Proc. 6201, 62012H

[16]   Primera-Pedrozo, O. M.; Soto-Feliciano, Y.; Pacheco-Londoño, L. C.; Hernandez-Rivera, S. P. *Sensing and Imaging: An International Journal.* 2008, 9, 27–40.

[17]   Primera-Pedrozo, O. M.; Pacheco-Londono, L. C.; Ruiz, O.; Ramírez, M.; Soto-Feliciano, Y. M.; De La Torre-Quintana, L. F.; Hernandez-Rivera, S.P. *Proceedings of SPIE.* 2005, 5778, 543-552.

[18]   Manrique-Bastidas, C. A.; Mina, N.; Castro, M. E.; Hernandez-Rivera, S.P. *Proceedings of SPIE.* 2005, 5794, 1358-1365.

[19]   Barreto-Cabán, M. A.; Pacheco-Londoño, L. C.; Ramírez, M. L.; Hernández-Rivera, S. P. *Proceedings of SPIE.* 2006, 6201, 620129-620139.

[20] Beeson, R; Skip, R. D., "Thermal inkjet technology – review and outlook. Rung Advanced Research Laboratory, Inkjet Business Unit, Hewlett-Packard Company, Corvallis Oregon, US.

[21] Lin-Vien, D.; Colthup, N. B.; Fateley, W. G.; Grasselli, J. G. The Handbook of Infrared and Raman Characteristic Frequencies of Organic Molecules. San Diego, CA: Academic Press.1991, pp. 179–189.

[22] White, G. M. *J. Forensic Sci.* 1992, 37, 652-656,

[23] Bellamy, A. J. *J. Forensic Sci.* 1999, 44, 603-608.

[24] Evans, H. K.; Tulleners, F. A. J.; Sanchez, B. L.; Rasmussen, C. A. *J. Forensic Sci.* 1986, 31, 1119-1125.

[25] Oxley, J. C.; Smith, J. L.; Shinde, K.; Moran, J. *Propellants, Explosives, Pyrotechnics.* 2005, 30 No. 2, 127-130.

[26] Oxley, J. C.; Smith, J. L.; Luo, W.; Brady, J. *Propellants Explos. Pyrotech.* 2009, 34, 539-543.

[27] Jattner, J.; Hoffmann, H. Handbook of Vibrational Spectroscopy. In: Chalmers, J. M., Griffiths, P. R., Eds. Wiley and Sons, Chichester, UK. Vol. 2, pp 1009-1027. 2002.

[28] Beebe, K.; Pell, R.; Beth, M. Chemometrics: A Practical Guide. John Wiley and Sons. N.Y. 1998

[29] OPUS 4.2 Version. User Manual. Bruker Optics: Billerica, MA. 2003.

[30] Skoog, D. A.; West, D. M.; James Holler, F.; and Crouch, S. R. Fundamentals of Analytical Chemistry. Belmont, CA: Thomson Brooks Cole. 2004.

[31] Thomsen, V.; Schatzlein, D.; and Mercuro, D. *Spectroscopy.* 2003, 18, 114.

[32] Stone, M.; Jonathan, P., *J. Chemometrics.* 1993, 7, 455-475.

[33] Johnson, R. A.; Wichern, D. W. Applied Multivariate Statistical Analysis. Prentice-Hall: Englewood Cliffs, NJ. 1992.

[34] Mardia, K. V.; Kent, J. T.; Bibbly, J. M. Multivariate Analysis; Academic Press: New York, 1979.

[35] Huberty, C. J. Applied Discriminant Analysis. Wiley–Interscience: NJ. 1994.

In: Fourier Transform Infrared Spectroscopy
Editor: Oliver J. Rees, pp. 101-116

ISBN: 978-1-61668-835-6
© 2010 Nova Science Publishers, Inc.

Chapter 4

APPLICATION OF CHEMOMETRIC MODELED FOURIER TRANSFORM INFRARED SPECTROMETRIC DATA FOR DIAGNOSTIC INVESTIGATION OF CANCER VIA TISSUE AND BLOOD ANALYSIS

Mohammadreza Khanmohammadi *
and Amir Bagheri Garmarudi
Chemistry Department, Faculty of Science, IKIU, Qazvin, Iran

ABSTRACT

Recently, Fourier transform infrared (FTIR) microspectroscopy has been employed to detect cancer tissues from normal ones. In most of these researches the infrared spectral characteristics have been analyzed by statistical tools to study variations in metabolites. It is effective to utilize chemometric techniques for reliable modeling and classification of human tissue samples from different cancers. Principal components analysis (PCA) and analysis of variance (ANOVA) techniques are used to make an initial decision about the feasibility of microspectroscopic analysis. Multivariate data analysis procedures based on linear discriminant analysis (LDA) and soft independent modeling of class analogy (SIMCA) are also performed to classify the obtained spectra. Studying the tissue samples has drawbacks such as difficult sampling, low speed sample preparation and very sensitive keeping conditions. Also sample contamination would affect the diagnosis results while the sampling procedure from human may help the malignancy to be more developed. Thus, blood samples from healthy people and those affected by cancer are also possible to be analyzed by ATR-FTIR microspectroscopy and the obtained data would be processed by aforementioned chemometric techniques. Initial modeling is performed to classify the blood samples from healthy people and malignant cases. Then basal cell carcinoma is investigated as the case study. The obtained results confirm that FTIR microspectroscopy in combination with chemometrics would be of high interest to be introduced as a diagnostic route in medical oncology.

---

* Tel: +982818371370; Fax: +982813780040; Email: mrkhanmohammadi@gmail.com

INTRODUCTION

Cancer is a leading cause of death worldwide: it accounted for 7.6 million deaths in 2005 and 7.9 million in 2007 (around 13% of all deaths). Lung, stomach, liver, colon and breast cancer cause the most cancer deaths each year. World Health Organization (WHO) in the beginning of its website intro mentions some meaningful and impressive facts about cancer: "Cancer affects everyone – the young and old, the rich and poor, men, women and children – and represents a tremendous burden on patients, families and societies. Cancer is one of the leading causes of death in the world, particularly in developing countries. Yet, many of these deaths can be avoided. Over 30% of all cancers can be prevented. Others can be detected early, treated and cured. Even with late stage cancer, the suffering of patients can be relieved with good palliative care." It has been proved that early detection of cancer greatly increases the chances for successful treatment. Researches in field of cancer diagnosis are promoted as intensive as the efforts for providing new effective therapies. It has been reported that about 30% of cancer deaths can be prevented. Cancer arises from a change in one single cell. The change may be started by external agents and inherited genetic factors. Deaths from cancer worldwide are projected to continue rising, with an estimated 12 million deaths in 2030. There are two major components of early detection of cancer: education to promote early diagnosis and screening. Recognizing possible warning signs of cancer and taking prompt action leads to early diagnosis. Increased awareness of possible warning signs of cancer, among physicians, nurses and other health care providers as well as among the general public, can have a great impact on the disease. Some early signs of cancer include lumps, sores that fail to heal, abnormal bleeding, persistent indigestion, and chronic hoarseness. Early diagnosis is particularly relevant for cancers of the breast, cervix, mouth, larynx, colon and rectum, and skin. Screening refers to the use of simple tests across a healthy population in order to identify individuals who have disease, but do not yet have symptoms. Examples include breast cancer screening using mammography and cervical cancer screening using cytology screening methods, including Pap smears. Screening programs should be undertaken only when their effectiveness has been demonstrated, when resources (personnel, equipment, etc.) are sufficient to cover nearly all of the target group, when facilities exist for confirming diagnoses and for treatment and follow-up of those with abnormal results, and when prevalence of the disease is high enough to justify the effort and costs of screening. The origin of cancer has been traced to alterations in the genetic composition of cells as a result of heredity, environment or a failure of normal cell growth patterns. Yet the biochemical changes occurring in different cancers are different [1]. There are several diagnostic approaches for cancer detection, but the biopsy sampling followed by professional pathological assessment is known as the "gold standard" and common procedure used to diagnose cancer. This standard histopathological procedure consists of tagging and visualizing the distribution and structure of cellular components in tissue sections using light microscopy. Although used in clinical routine, this common approach is not free of pitfalls and is associated with environmental contaminations by agents used for sample preparation [2,3]. It has been mentioned in literature that the biopsy based approach is a time-consuming method and is based on the pathologist's professional capabilities. Although diagnostic procedures have been improved tremendously during recent decades, they are established as highly expensive techniques. Thus there are two main concerns in this field: reducing the effect of pathologist's experience on obtained results and investigating new diagnostic strategies which are rapid, accurate, precise, reliable and also economically feasible. One of the newest ideas in this field is to apply infrared spectroscopy as an inspection method in diagnostic approaches. Describing the biochemical structure of our body, it is generally composed of water, lipids, nucleic acids, proteins and carbohydrates. It has been mentioned in the research repots and literature that any disease would be the result of changes in the body biochemical. Every biochemical object in a bio-system would demonstrate a reflex to infrared emission as a function of intra-molecular bands' vibrations [4-6]. Infrared spectroscopy has been successfully established as a reliable method, applicable in investigation of biological structures and molecules. Furthermore the biospectroscopy research trends have been focused on more detailed analysis of cells and tissues. Although it is a very difficult job to provide any detailed and specific information of molecular level structures and the complexity of infrared spectra is more intense in such systems. Thus it is more effective to assess the spectra for pattern recognition [7,8]. There are several advantages mentioned for infrared spectroscopy such as:

1.   Rapid data recording, as the spectral data is collected and interpreted within minutes
2.   No sample preparation step is required prior to spectral analysis or it requires minimal sample preparation
3.   Non-destructive analysis which enables the researcher to perform some further diagnostic tests on the IR investigated samples
4.   Reliability due to the possibility for constructing a large model of known samples as a data bank for further comparison tests
5.   Possibility for application of statistical techniques in construction of classification models (Chemometrics)
6.   Reducing the probable misdiagnosis which is common while some other approaches such as pathologic investigations are used
7.   Analysis is not limited in accordance with state of samples. Spectra can be obtained from liquids, solids (pellets, powders, films and tissues), slurries and suspensions.

In this chapter, the main trends in field of cancer diagnosis by infrared spectroscopy are reported. Also, a novel application of this technique based on blood analysis is discussed.

VIBRATIONAL SPECTROSCOPIC FEATURES OF CELL BIOCHEMICALS

Infrared spectroscopic analysis of biological samples is faced to investigate the biochemical characteristics of the bio system. There are several biochemicals in a tissue or bio-fluid samples. The main are protein, nucleic acids, carbohydrates and lipids. The proper functioning of proteins is related to its structural variations. Infrared spectroscopy provides information on the secondary structure of proteins, ligand interactions and folding [9]. The main spectral observed signals in relation with protein structures are due to amide A, amide B and amides I–VII. The C=O stretching band of amide I (related to peptide backbone) is often the most intensive one. This band is usually observed in the 1600–1700 cm^{-1} region and has been used the most for structural studies due to its high sensitivity to small changes in molecular geometry and hydrogen bonding of the peptide group. Also a coupling of N-H in-plane bending and C-N stretching would form the amide II band in the 1480–1570 cm^{-1} region. It is mentionable that amide II band is not as useful as amide I because of its les sensitivity and also probable interference with amino acid side chains' vibrational signals. In addition with protein structural vibrations, the role of DNA and RNA is also important as they carry information that determines proteins' structure. The main spectral features of nucleic acids in infrared region are located in 1750–1600, 1230, 1100 cm^{-1} which are due to in-plane double-bond vibrations of the bases, asymmetric and symmetric phosphate stretching respectively. Lipid as the major component of cell walls keeping biological media organized in their necessary compartments and carbohydrates as the main energy source of cells are also important while a biological system is studied. These biochemicals' main spectral characteristics are located in 1800-900 cm^{-1} spectral region [10-12]. It is noticeable that the case of tissue sample analysis is too much more complex in comparison with this single component biochemical vibration feature explanation. In a bio system, the probable effect of each structure on the other ones, amplifying or reducing interferences and some shifts are also present which would make a huge amount of complex information to be resolved. In such a case pattern recognition strategy is helpful.

THEORETICAL DESCRIPTION OF APPLIED CHEMOMETRIC TECHNIQUES

In order to prove the most appropriate designation in a quantitative or qualitative analysis, chemometric disciplines are applied which use mathematical and statistical routes for data processing, providing the optimal experimental situation. Chemometrics would also provide the maximum chemical information by analyzing chemical data. The data processing strategy is applied in many fields of chemical research e.g. quantitative analysis, qualitative analysis, structure-activity or structure-property relation, etc. but one of the first and most publicized success stories is pattern recognition. Much chemical researches involve using data to determine patterns and thus the chemometric techniques (both unsupervised and supervised) are useful for this aim. During the chemometrics developments, multivariate statisticians have introduced many discriminating functions, some of direct relevance to chemists. Considering application of a spectrochemical method such as FTIR spectroscopy to determine whether a sample tissue is cancerous or not, a mathematical model is set up, being calibrated by some known samples. Finally by using the proposed model, the diagnostic situation of an unknown sample can be predicted. One of the initial requirements of chemometric data processing for pattern recognition and classification goals is to assess the possibility of classification in the data set. Analysis of variance (ANOVA) is a useful technique for comparing two or main classes, treatments or methods. It is similar to regression in that it is used to investigate and model the relationship between a response variable and one or more independent variables. However, analysis of variance differs from regression in two ways: the independent variables are qualitative (categorical), and no assumption is made about the nature of the relationship. In the other words, the model does not include coefficients for variables. ANOVA is applied prior to any other data processing in order to evaluate the possibility for classification tissue sample spectra. The next step is to start the main data processing procedure. The first idea is to apply principal component analysis (PCA). PCA is a common method which reduces the complex spectral data into much fewer dimensions using principal components (eigenvectors) and scores. Using this method, the extent to which IR features can differentiate normal and cancer tissues is investigated. The objective of this section was to conduct the diagnostic test, utilizing the simplified spectral data, treated by PCA. Another useful approach for pattern recognition and classification aims is linear discriminant analysis (LDA). This technique is a powerful tool which builds discrimination rules that are used to identify the situation of unknown samples. LDA searches for the variables containing the greatest interclass variance and the smallest intraclass variance, and constructs a linear combination of the variables to discriminate between the classes. The rule is constructed with training set of samples, and further tested with the test set. In this research, we tried to study colon LDA, in order to improve the reliability of the interpretation of the data [13,14]. Soft independent modeling of class analogy (SIMCA) is a technique, known as unsupervised methodology. PCA is a pre-processing step in SIMCA in order to retain the sufficient number of principal components which must also be optimized. Retention of too few components can distort the signal or information content contained in the model about the class, whereas retention of too many principal components diminishes the signal-to-noise ratio. This optimization is usually performed by cross-validation. In order to analyze an unknown sample, the residual variance of an unknown sample is compared to the average residual variance of those samples that make up the class. It is possible to obtain a direct measure of the similarity of the unknown to the class. This comparison is also a measure of the goodness of fit of the sample to a particular PC model. Often, the F-statistic is used to compare the residual variance of a sample with the mean residual variance of the class [15-17]. An advantage of SIMCA is that an unknown is only assigned to the class for which it has a high probability. If the residual variance of a sample exceeds the upper limit for every modelled class in the data set, the sample would not be assigned to any of the classes because it is either an outlier or comes from a class that is not represented in the data set. SIMCA can work with as few as 10 samples per class, and there is no restriction on the number of measurement variables, which is an important consideration, because the number of measurement variables often exceeds the number of samples in chemical studies.

ANALYSIS OF COLON CANCER TISSUE SAMPLES

Colon cancer has allocated one of the highest mortality rates. It is the 3rd leading class of cancer causing increased mortality in developed countries. Every year in the US, more that 150000 colon cancer cases are diagnosed while about 65000 deaths would occur from colorectal cancer. If colorectal cancer is detected at an early stage, then the 5-year relative survival rate is 90%; however only about 37% of colorectal cancers are diagnosed at early stages. The main indicator for the initiation of malignancy is abnormal cell proliferation. Various techniques have been utilized to identify abnormal crypt proliferation in colon tissues. The presence of certain molecular markers in the crypt has been used to detect abnormal proliferation [18]. The FTIR microspectroscopy is capable of detecting malignancy among colon cancer patients using biopsy tissue samples. Studying the colon cancer cell lines demonstrate several of the spectroscopic features displayed by colon cancer tissues. Cell lines have been as an appropriate experimental system both to study the origin of the spectroscopic changes observed hitherto and to apply the IR technology to other aspects of cancer biology. The splitting pattern has been observed in 1150-1000 cm^{-1} spectral region for normal cells and are absent in polyp and malignant cells. However, an extensive set of spectroscopic changes are common between the malignant tissue and colon adenocarcinoma cell lines. FTIR spectra of normal and malignant cells from tissue samples of different colon cancer patients show the presence of two families: normal and abnormal cells. Some infrared parameters of malignant colonic tissue obtained at atmospheric pressure are significantly different from those of normal colonic tissue e.g. a 2–3 cm^{-1} difference in the frequency of the symmetric phosphate stretching band of the nucleic acids [19]. Most of the reports in the field of cancer detection by FTIR spectroscopy are based on experimental investigation of isolated signal intensities or signal position shift and related signals' intensity ratio. The lack of a statistical-mathematical approach for improving the reliability of diagnostic results is observed. In order to overcome this problem chemometric techniques are applied while research is performed in order to combine the statistical skills with chemical ones and gain as insurable as possible results. Multivariate data analysis is one of these techniques that have been presented by several researchers [20]. The obtained analytical data are processed by chemometric procedures in order to improve the capabilities in prediction of unknown samples. Multivariate methods are powerful in the evaluation of complex spectral data corresponding to environmental and biological samples. Advanced chemometric techniques i.e. PCA, LDA, CA and ANN are applied in many researches; They could successfully explain the complex data in classification, pattern recognition and also quantitative analysis [21-24].

Samples, Instruments and Software

Colonic tissue specimens were obtained by the pathology division in Shahid Rajaee Hospital Qazvin University of Medical Science, Qazvin, Iran (Totally 78 samples from 42 males and 36 females). It is noticeable that total number of samples was selected according to the cases which were in clinical relation with the hospital during the investigation time period. The patients' age range was 59-91 years and they had all underwent pathologic diagnostic test procedures. The ethical guidelines were observed properly and permission was obtained from all subjects. Patients were informed about the research goals and participated in the sampling voluntarily. All the samples were investigated diagnostically by the pathology department of QUMS and the evaluation was repeated by second and third look. Thus the reliable resulted samples were chosen to be analyzed by FTIR spectrometry, while the pathology gold standard method's results were supposed as the reference. All FTIR microspectroscopic studies were performed using a Nicolet® spectrometer equipped with a microscope (Spectratech®; Colorado, USA) at mid-IR region and ZnSe ATR accessory (avoiding the strong absorption of ordinary glass slides mid-IR region), using liquid nitrogen cooled mercury cadmium telluride (MCT) detector. During spectra acquisition, the measured sites' area was about 50×50 μm^2 at most, which contains only few cells. Number of 49 scans and 8 cm^{-1} resolution was set as the optimum condition. It is also noticeable that for each sample at least 3 spectra were recorded in order to monitor the repeatability. Baseline correction for all the spectra was done by the software, finding the best fit to the background using an automatically polynomial function. In order to avoid the effect of contamination on spectral studies, all tissue areas in microspectroscopy studies were selected by microscopic inspection in parallel with detailed pathological analysis of the tissue architecture. The microscope was also equipped with a visible region camera. All spectral studies were performed in 4000-400 cm^{-1} wavenumber region. In order to apply the chemometrics methods, ANOVA, PCA, LDA and SIMCA techniques were performed by MATLAB® version 7.1 software (The MathWorks Inc., MA, USA). Spectroscopic data were processed in selected spectral region by this software.

Comparison of Spectral Data for Colon Cancer Tissue Samples

In order to compare the signals in the FTIR spectra of the tissue samples, the baselines of the spectra were initially corrected and normalized by intensity at 1800 cm^{-1}. Although the protein structural change between normal and cancer tissue samples is significant, there are several spectral features in 1800-900 cm^{-1} region which are due to different biomolecules. Two intensive signals at 1660 cm^{-1} and 1540 cm^{-1} which are due to amide I and amide II respectively, are indicative of protein structure. Cancerous colonic tissues would demonstrate a systematic decrease in total carbohydrate and phosphate content of cells. Although, the biological origin of the spectral differences are difficult to be fully addressed. Phosphate stretching modes could be useful as infrared spectroscopic markers to discriminate between spectra of normal and cancerous colonic tissues. Spectral features due to stretching vibrations of PO_2^- from the nucleic acids and the C–O stretching vibrations from the carbohydrate residues in the glycogen (or collagen) are found in 1240 and 1250 cm^{-1}. Some other bands are symmetric $\delta[C(CH_3)_2]$ at 1400 cm^{-1}, CH_2 wagging at 1335 cm^{-1}, asymmetric $\delta[CH_3]$ at 1455 cm^{-1}. Also, it is important to identify the special patterns of vibrational modes characteristic to various states of malignancy. Investigating typical infrared absorption spectrum of normal and cancerous human colon tissues demonstrates the main differences in 1800-900 cm^{-1} spectral region. It is noticeable that some signals appearing in this region are not as intensive as aforementioned ones but their presence is important for further classification processes. Some of these are: υ(C-O), υ(C-O-H), υ(C-C) all around 1160 cm^{-1}; $\upsilon(PO_2^-)$ at 1080 cm^{-1}; $\upsilon(CH2OH)$, υ(C-O) and C-O bending at 1040 cm^{-1}. The bands at 1025 and 1045 cm^{-1} are responsible for the vibrational modes of –CH_2OH groups and the C–O stretching coupled C–O bending of the C–OH groups. On the other hand, lipids, polysaccharides, and other biochemicals would also have many signals in the 1800-900 cm^{-1} spectral region [10,12]. However, these spectral differences can provide information that is useful for the discrimination between normal and cancer, especially if a multivariate chemometric scheme is used in combination with the visual differences. Considering some of the spectral differences would not be a useful idea for discrimination between normal and cancer tissues. According to the complex structure of the spectra, it is important to consider the total region for better diagnosis. A key aim of experimentation is to ask how significant a factor is in colon cancer spectra. Thus it was decided to perform all the chemometrics procedures for total spectral region (1800-900 cm^{-1}).

RESULTS AND DISCUSSIONS

ANOVA and PCA Results

As mentioned previously, prior to any data processing, ANOVA was performed to be ensured of classification possibility for colon cancer tissue and healthy ones. The obtained box plot confirmed the formation of 2 classes due to different calculated variance for each class. PCA was utilized to simplify the data interpretation by reducing the variable dimension of the spectral data matrix down to minimal number of orthogonal principal components (PC). It was applied into data set containing 60 samples. Since the new PCs summarize most of the information in the data, if we can interpret these, then we have greatly simplified the analysis. These PCs (t_1, t_2,..., t_A) are simply linear combinations of the variables that summarize nearly all of the variance in the original data matrix. The mathematical equation in matrices is:

$$X (N \times K) = T (N \times A) P^T (A \times K) + E$$

where T is a matrix with only A ($<$ N) columns containing the PCs. The loading matrix P shows how these PCs are related to the absorbnce at the K wavenumbers, and each row of P^T can be interpreted as a type of spectrum that defines the corresponding new variables. E is the residual of the spectral data matrix e.g. noises.

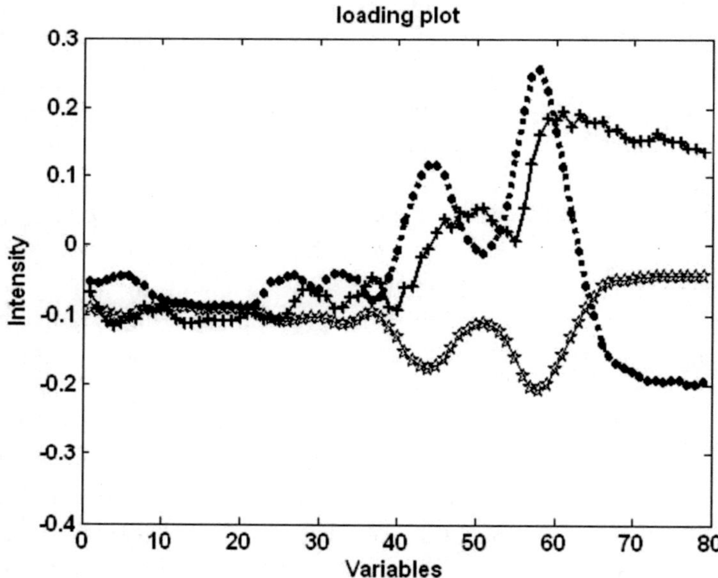

Figure 1. PCA loading plot for colon tissue FTIR spectra demonstrating the optimal number of PC's.

Captured variance for 3 first PC'S was 99.7 % that illustrate an acceptable value for this method. The loading plot demonstrated that first two PCs have an important role in modeling and 3rd one is not helpful enough (figure 1). Thus the first two PCs were selected for classification of samples. While the PCA obtained score plot illustrated a regional separation between normal and cancer case spectra, this ability was not clear in some regions of the score plot, and it was decided to apply more powerful techniques.

LDA Results

LDA was introduced as a linear classification strategy in the experimental section. It is a linear combination for a discriminant analysis, derived as:

$$Y = z_1 X_1 + z_2 X_2 + z_3 X_3 + \cdots + z_n X_n$$

where Y is the discriminant score, z_i is the discriminant weight for independent variable i and X_i is the independent variable i; in our study, the independent variable corresponds to the spectral absorbance in a specific wavenumber. Each z_i is set to maximize the between-group variance of y (variance between cancer set and normal set) relative to the within-group variance of y (y variance within cancer group and the variance within normal group). A critical Z-value is also defined by and exponential equation in accordance with both classes' members and this critical Z is used as the boundary value. LDA was applied to discriminate the two classes (normal and cancer). The obtained FTIR spectra were statistically analyzed to determine differences between malignant status and normal status. The results obtained in PCA step were used prior to LDA. The training set for the LDA calibration model was consisted of 30 tissue spectra (15 healthy and 15 cancer). The same as every other analysis method, the next step was to evaluate the power of optimized model in prediction of unknown samples. The remaining 48 colon tissue spectra consisting of 29 healthy and 19 cancer cases (not included in the training set) were analyzed in order to provide a more generalize able model. Comparison between the LDA obtained results with those of standard pathologic reports illustrated that 46 samples have been predicted correctly, while 2 healthy tissue spectra were predicted as cancer cases wrongly. Thus the accuracy was 95.8%. LDA's sensitivity and specificity were determined 100 and 93.1%, respectively.

SIMCA Results

Again the PCA results were used to start the SIMCA. Utilizing cross-validation for each class (healthy and cancer) in accordance with minimum value of Root Mean Square of Cross Validation (RMSECV) it was tried to examine the SIMCA model. It is necessary the form training or calibration model and then perform the model reliability check procedure. Appling PCA, Q as a residual error for prediction the model was 7×10^{-4} and 1.4×10^{-3} for cancerous and healthy class in calibration set respectively.

Mathematically s_0^k as the standard deviation of the training set for class k is calculated as:

$$s_0^k = \left[\sum_{j=1}^{m} \sum_{i=1}^{n_k} \frac{\left(e_{ij}^k\right)^2}{\left(n_k - p_k - 1\right)\left(m - p_k\right)} \right]^{1/2}$$

where n_k is the number of objects, p_k is the number of significant principal components in class k, m is the number of variables, and e_{ij} is the residual. The standard deviation of the unknown spectrum, s_u^k, is then calculated using e_{uj} value:

$$s_u^k = \left[\sum_{j=1}^{m} \frac{\left(e_{uj}^k\right)^2}{(m - p_k)} \right]^{1/2}$$

The same as LDA a critical value is calculated F-distribution with the 95% confidence interval (F-value) to be set as the boundary level. If the F-value of an unknown sample is larger than the critical F-value at a given level of significance, one can conclude that the distance (u) from class k is significantly larger than that of the class k as a whole and it is determined to be excluded from class K. If the distance is not significantly larger than F-value, the sample is determined to be a member of class K. We need an analytical way to find the most extreme points in the data. T^2 or Hotelling's parameter indicates a statistical measure of the multivariate distance of each sample from the centre of the data set. The comparison plot of Q versus T^2, in figure 2 indicates the ability of SIMCA calibration model which was 18 samples (9 healthy and 9 cancer cases).

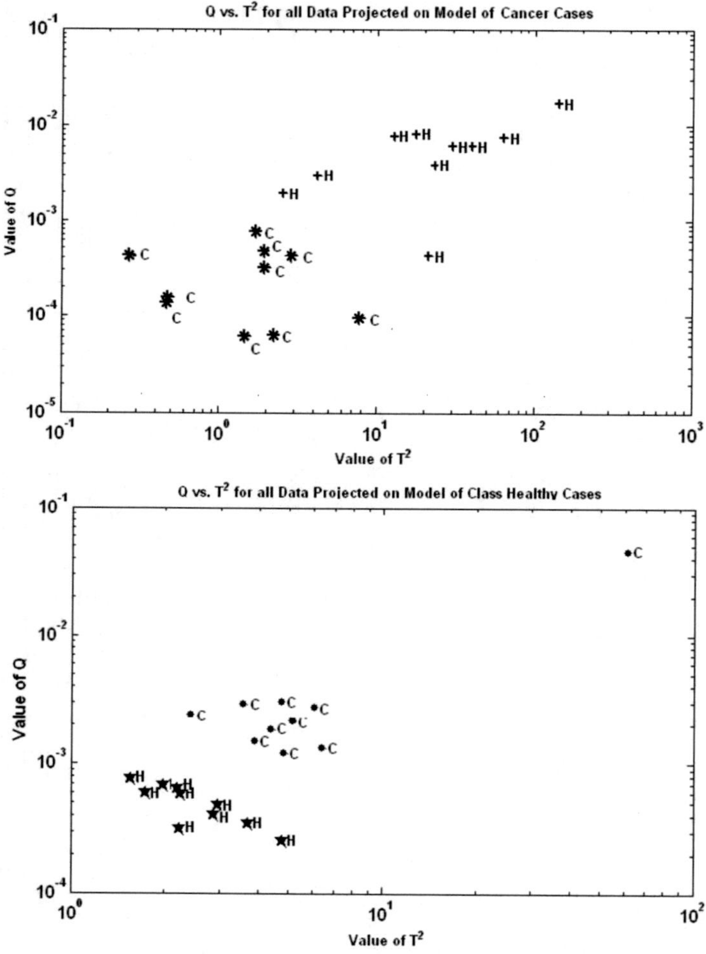

Figure 2. Schematic presentation of SIMCA training results for colon tissue samples' FTIR spectra.

Finally, 60 unknown samples (26 normal and 34 cancer) were predicted by SIMCA model and 4 cancer tissue samples were predicted to be normal wrongly. The accuracy, sensitivity and specificity of the model were 93.3 and 88.2 and 100% respectively. The final step was to investigate the effect of H and E staining on the results of tissue classification by ATR-FTIR microspectroscopy. A non-stained slide of each sample was also studied in the same procedure of FTIR microspectrometry being processed by all the aforementioned chemometrics techniques. The obtained results were completely the same as stained samples and even the wrongly predicted samples were from the same patients in both models. Thus it was concluded that is possible to study the unstained tissue samples for diagnosis of colon cancer while selecting the tissue area to be investigated is easier in stained tissues. This advantage would enable the specialists to omit the H and E staining step which is somehow making some difficulties.

ANALYSIS OF BLOOD SAMPLES FOR CANCER DETECTION

While the main research activities for cancer diagnosis by FTIR spectrometry have been focused on tissue analysis, it is noticeable that tissue samples are normally very heterogeneous, comprising several components and different cell types. The preparation procedure of biopsy samples consists of 4 main sub-procedures: fixation, embedding, sectioning and mounting on a glass slide and finally staining. Although the whole procedure is time consuming and difficult to operate, fixation and embedding process are dangerous, as the treatment may distort the cell structure and immersion of tissues in organic solvents, such as xylene, dissolves the tissue lipids. There are also some other drawbacks in studying the tissue samples; the main ones are:

1- low speed sample preparation
2- difficult sampling due to troubles in accessing the body organ from which the sample is required
3- very sensitive keeping conditions for the sample tissue prior to analysis
4- most of the biopsy operations are also invasive and may cause spread of the disease

Thus it seems of high interest or even necessary to propose some new analytes which could be inreoduced as substitute for tissue samples, while easier providing and analysis is possible and reliable results are achieved. Besides the application of FTIR spectrometry in diagnostic approaches using tissue samples as the analyte, its role in the diagnostic aspects involving body fluids' analysis is gaining importance in the last few years. As the malignant cells' adhesion strength is lower than healthy ones, they would be separated easily and also travelling via body fluids from the cancerous section to healthy ones. This process could also be invasive and cause metastasis. Blood is a main travelling media for these cancerous cells and an idea is to replace blood sample instead of tissue samples in cancer diagnostic studies, as blood is more homogeneous and the sample preparation procedure is very simple while blood is studied. In the proposed idea for blood analysis and in the first step, blood samples from healthy people and those with cancer where tried to discriminated by LDA. The basal cell carcinoma (BCC) as a specific cancer was selected and the blood samples from healthy people and those with BCC were analyzed.

Samples, Instruments and Software

There was a sample preparation step which could also be mentioned as an advantage of blood analysis. A 2 ml volume sample blood sample was provided and kept at 25 °C in EDTA vials (as anticoagulant agent) , being analyzed till 20 hours after bleeding. In general discrimination procedure 43 blood sample (24 cancerous and 19 healthy) were analyzed. The cancer cases from colon, prostate, breast, skin, etc. For diagnosis of BCC, 72 sample (32 healthy and 40 cancerous) were analyzed. The patients who participated in blood sampling were 55-72 years old and the ethical guidelines were all spotted the same as tissue analysis section. Patients must be selected from those who are not treated by chemotherapy, because the blood samples must be pure, otherwise the comparison between healthy and cancerous samples is failed.

Spectrometric Findings

The main problem in the spectral analysis in mid-IR region was the absorbance of water which consists about 90 percent of blood. The water related signals would demonstrate interference, covering all the useful spectral features. Water was set as background spectrum, avoiding this problem. Red blood cells (RBC: erythrocyte), which carry respiratory gases and give blood its red color because they contain haemoglobin, white blood cells (WBC: leukocytes), which fight disease, and platelets (thrombocytes), cell fragments which play an important part in the clotting of the blood are the main cellular constituents of blood. These cells are suspended in fluid medium known as plasma. The spectral signals and bands observed due to these ingredients are similar to those described in the previous sections as "Vibrational spectroscopic features of cell biochemicals". Several functional groups would affect the IR spectra of blood samples via their finger print signals. According to the studies in accordance with biomarkers and tumor biology, the main biomolecules to be investigated are proteins. It is noticeable that the most intensive signals due to protein content of blood are in 900-1800 cm^{-1} spectral region which was selected for analysis. The intensity of a signal is directly dependent to the quantity of the functional group (amide groups here). Comparing the different spectral regions of IR spectrum the bands located at 1550 and 1630–1680 cm^{-1} are the most intensive ones. These are both observed in cancer and healthy samples' spectra. The band at 1630–1680 cm^{-1} is due to the C=O stretching of amide group. Absorbance bands at 1600–1720 cm^{-1} in the FTIR spectra of human blood are related to the amide I vibrational modes of blood proteins and exhibit a high sensitivity to conformational changes in the secondary structure. The RCONHR group has also a bending signal located at 1620–1640 cm^{-1}. In many situations, these two regions would overlap and their identification is impossible. Of course as the main aim is to make a pattern recognition approach, these probable interferences would not influence the final results. In this region, there is also a band at 1550 cm^{-1} due to the C–N stretching and N-H bending of R-CONR$_2$. The amide II band mainly stems from the C–N stretching and C–N–H bending vibrations weakly coupled to the C=O stretching. The main signals of amide II appear in the 1539–1550 cm^{-1} region, due to a-helix. However, this region is not as sensitive to conformational changes as reported for the amide I bands. As detailed, most of the intensive bands related to biomolecules of blood constituents are in 1800–900 cm^1 region and this region was selected to be investigated.

Results and Discussion

LDA Results for General Discrimination

As mentioned previously, the first study was to investigate to probable possibility for discrimination between IR spectra of blood samples from healthy people and those affected by variety of cancers. Thirty three samples (14 cancer and 19 healthy) were introduced to the LDA model as a training set in order to form a calibration model (figure 3).

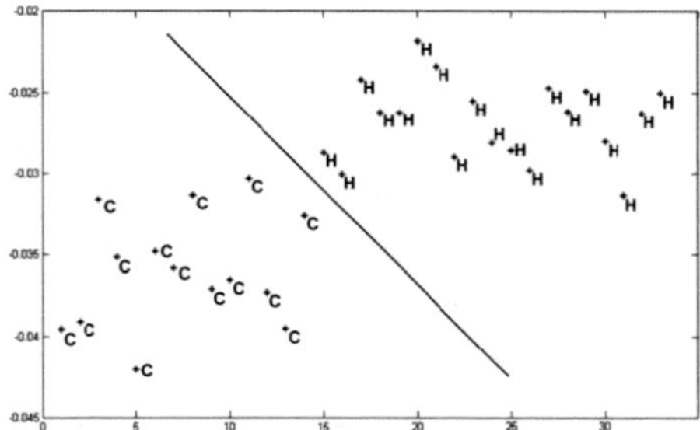

Figure 3. Validation of LDA calibration model for diagnosis of cancer via blood IR spectra

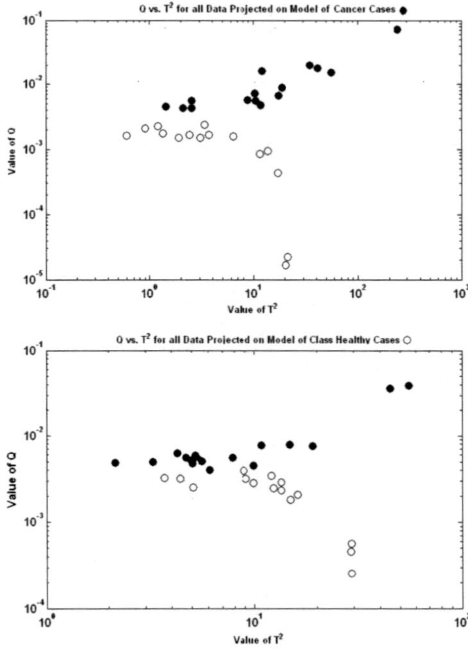

Figure 4. SIMCA validation scheme for diagnosis of BCC via infrared spectrometric analysis of blood samples

Then 10 unknown samples were predicted (5 of each class). Nine samples were diagnosed correctly and 1 healthy was misdiagnosed as cancer case. Thus the accuracy of the proposed LDA-FTIR for general diagnostic method was about 90% which could be acceptable for initial study while the amount of samples was low (only 43 samples). It could be desired that large data bases of blood samples would enables the FTIR to be introduced as a robust diagnostic method.

SIMCA Results for BCC Diagnosis

In SIMCA data processing for BCC diagnosis, healthy and cancer classes were defined and each class was independently modeled using PCA. It is mentionable that each class could be described by specific number of PC's. According to 95% of interval confidence, the residual error of model was about 0.003 for both classes. In validation step, 30 samples (15 of each class) were introduced to the formed model in order to calibrate the classification set up. Then 42 unknown samples (17 normal and 25 cancer cases) were predicted by the model and the obtained results were compared with those of clinical pathologic inspections. All the 17 normal samples were predicted correctly and among the 25 cancer samples, 24 samples were diagnosed as cancer case while 1 sample was predicted as a normal case. Accuracy, sensitivity and specificity of the proposed diagnosis method are 97.6, 94.4 and 100% respectively.

CONCLUDING REMARKS

While the application of infrared spectrometry for diagnostic analysis of tissue sample in cancer research and clinical oncology is of high interest, proposing some novel methods which would introduce new analytical or sample preparation techniques would be of high interest for health related systems. Introducing blood analysis as a non-invasive, rapid, reliable and robust idea for cancer diagnosis would be a new edge of knowledge in this field and it seems that more developed researches are required in order to make this idea as a useful and common approach.

REFERENCES

[1]   www.who.org ; accessed during June-September 2009.

[2]   Mordechai, S; Sahu, RK; Hammody, Z; Mark, S; Kantarovich, K; Guterman, H; Podshyvalov, A; Goldstein, J; Argov, S. *Journal of Microscopy*, Possible common biomarkers from FTIR microspectroscopy of cervical cancer and melanoma, 2004, 215, 8691.

[3]   Fabian, H; Lasch, P; Boese, M; Haensch, W. *Journal of Molecular Structure*, Infrared microspectroscopic imaging of benign breast tumor tissue sections, 661, 411–417.

[4]   Pancoska, P; Kubelka, J; Keiderling, T.A. *Applied Spectroscopy*, Novel Use of a Static Modification of Two-Dimensional Correlation Analysis. Part I: Comparison of the Secondary Structure Sensitivity of Electronic Circular Dichroism, FT-IR, and Raman Spectra of Proteins, 1999, 53, 655-661.

[5]   Liquier, J; Taillandier, E. Infrared Spectroscopy of Nucleic Acids, In: H.H. Mantsch and D. Chapman, editors. *Infrared Spectroscopy of Biomolecules*. New York: Wiley-Liss; 1996; 131.

[6]   Lewis, R.N.A.H.; Mcelhaney, R.N. Fourier Transform Infrared Spectroscopy in the Studyof Hydrated Lipids and Lipid Bilayer Membranes, In: H.H. Mantsch and D. Chapman,editors. *Infrared Spectroscopy of Biomolecules*. New York: Wiley-Liss; 1996; 1597.

[7]   Chalmers, J.M. and Griffiths, P.R., editors. *Handbook of Vibrational Spectroscopy*, vol. 5,Chichester, Wiley, 20028- Moss, D.A.; Keese, M.; Pepperkok, R. *Vibrational Spectroscopy*, IR icrospectroscopy oflive cells, 2005, 38, 185–191.

[8]   Dukor, RK. Vibrational Spectroscopy in the Detection of Cancer. In: J.M. Chalmers and P.R. Griffiths, editors. *Encyclopedia or vibrational spectroscopy*. New York: John Wiley;2002.

[9]   Pisani, P.; Parkin, D.M.; Bray, F.; Ferlay, J. *International Journal of Cancer*, Estimates of the worldwide mortality from 25 cancers in 1990. 1999, 83, 18–29.

[10]   Diem, M. Introduction to modern vibrational spectroscopy. New York: Wiley; 1993.

[11]   Krupnik, E.; Jackson, M.; Bird, R.P.; Smith, I.C.P.; Mantsch HH. *SPIE 1998*, Investigation into the infrared spectroscopic characteristics of normal and malignant colonic epithelium. 1998, 3257, 307–310.

[12]   Khanmohammadi, M.; Nasiri, R.; Ghasemi, K.; Samani, S.; Bagheri Garmarudi, A.*Journal of Cancer Research and Clinical Oncology*. Diagnosis of basal cell carcinoma byinfrared spectroscopy of whole blood samples applying soft independent modeling class analogy. 2007, 133, 1001-1010.

[13]   Brereton, R.G. Chemometrics: Data Analysis for the Laboratory and Chemical Plant.London: John Wiley and Sons; 2003.

[14]   Wold, S.; Albano, C.; Dunn, W.J.; Edlund, U.; Esbensen, K.; Hellberg, S.; Johansson,E.; Lindberg, W.; Sjostrom, M. Multivariate Data Analysis in Chemistry. In: B.R. Kowalskiand D. Reidel, editors. *Chemometrics, Mathematics and Statistics in Chemistry*. Dordrecht: Publishing Company, 1984.

[15]   Lavine, B.K.; Jurs, P.C.; Henry, D.R. *Journal of Chemometrics*. Chance Classificationsby Nonparametric Linear Discriminant Functions. 1988, 2, 1–10.

[16]   Khanmohammadi, M.; Bagheri Garmarudi, A.; Ghasemi, K.; Kazemi Jaliseh, H.;Kaviani, A. *Medical Oncology*, Diagnosis of colon cancer by attenuated total reflectancefourier transform infrared microspectroscopy and soft independent modeling of class analogy. 2008, 26, 292-297.

[17]   Potten, C.S.; Kellett, M.; Roberts, S.A.; Rew, D.A.; Wilson, G.D. *British MedicalJournal*. Measurement of in vivo proliferation in human colorectal mucosa usingbromodeoxyuridine. 1992;33:71–8.

[18]   Chen, Y.J.; Hsieh, Y.W.; Cheng, Y.D.; Liao, C.C. *Chang Gung Medical Journal*, Study on the secondary structure of protein in amide I band from human colon cancer tissue byFourier-transform infrared spectroscopy. 2001, 24, 541-546.

[19]   Khanmohammadi, M.; Ansari, M.A.; Bagheri Garmarudi, A.; Hassanzadeh, G.; Garoosi, G. *Cancer Investigation*, Cancer diagnosis by discrimination between normal and malignant human blood samples using Attenuated Total Reflectance - Fourier transform infraredspectroscopy. 2007, 25, 397-404.

[20] Morrison, D.F. Multivariate Statistical Methods. McGraw-Hill International: Auckland; 1978.

[21] Mardia, K.V.; Kent, J.T.; Bibby, J.M. Multivariate Analysis. Academic Press: London; 1989.

[22] Deming, S.N.; Morgan, S.L. Experimental Design: A Chemometric Approach. Elsevier: Amsterdam;1993.

[23] Wackernagel, H. Multivariate Geostatistic: An Introduction with Application, 3rd edition,Springer: Heidelberg, 2003.

In: Fourier Transform Infrared Spectroscopy
Editor: Oliver J. Rees, pp. 117-137

ISBN: 978-1-61668-835-6
© 2010 Nova Science Publishers, Inc.

Chapter 5

SIMULATOR PHOTOREACTOR FOR THE CHEMICAL CHARACTERIZATION OF DIESEL EXHAUST EMISSIONS

E. Borrás and *L.A Tortajada*

Fundación Centro de Estudios Ambientales del Mediterráneo (CEAM),
Valencia, Spain
Universidad Politécnica de Valencia, Valencia, Spain.

ABSTRACT

The study of diesel engine exhaust emissions is important due to their impact on atmospheric chemistry and air pollution. Although information on the general nature of diesel emissions is widely available, the gas and particulate phase characterization is still limited. Is known that hundreds of compounds are emitted from combustion and that they depend on type of engine, fuel composition, catalyst and engine conditions. In particular, the effects of diesel reformulation and engine operating parameters have an important impact on the gas and the particulate phase composition. Moreover, the primary diesel emissions are transformed following atmospheric degradation reactions (photolysis, photo-oxidation processes, darkness chemistry based on NO_3 and ozone). These conversions depend on initial exhaust composition, concentration of oxidants, sunlight intensity and atmospheric conditions.

In this chapter, a revision of main atmospheric transformations of diesel exhaust emissions has been performed. A comprehensive description of the monitoring strategies - dilution tunnels, on-road measurements, photoreactors, etc - has been also included. In this sense, the use of large photoreactors allows a better monitoring of different primary and secondary compounds, since precise conditions can be reproduced, isolated both from other pollution sources as other atmospheric processes. In the following section, advantages and limitations of this type of facilities have been discussed. For this, a complete description of the basic elements, the analytical instrumentation and experiment protocol has been performed. Finally, a state-of-art about diesel exhaust research in photoreactors has been also reviewed.

---

* Fundación Centro de Estudios Ambientales del Mediterráneo (CEAM), 46980 Paterna, Valencia, Spain

An example of case study - PAH experiments at European Photo-reactor (EUPHORE) - has been included. These condensed compounds are important due to their high toxicity and their correlation with gaseous compounds, chemical composition of fuels and engine operating conditions. In conclusion, the kind of information provided by photoreactors, contributes for the development of air quality programs and for the design of new alternative engines and fuels.

INTRODUCTION

In spite of substantial improvements in engine technology, a wide range of gaseous and particulate phase organic and inorganic compounds are still associated with diesel exhaust emissions [1,2]. The composition of diesel emissions contrasts strongly with the typical chemical composition of the atmosphere [3]. To assess pollutant impacts and the consequent changes to climate and the environment at the surface of the Earth, there is a need to understand the processes occurring in exhaust emissions and their coupling with the natural constituents of the atmospheric system.

Exhaust is produced in the combustion chamber of diesel engines and is emitted in the form of a plume of gases and particulate matter from the engine pipe. Once released into the troposphere, these compounds are dispersed, transported and transformed by various time-dependent physical and chemical processes [4]. In detail, pollutants spread in a three-dimensional flow immediately after their generation, mixing with other atmospheric components (Figure 1). The dilution factor is a complex function of many parameters depending on the atmospheric conditions. Simultaneously, some species react to form secondary pollutants through several chemical transformations. The products of these reactions are generally at a higher oxidation state, thus more polar and better soluble in water. These secondary pollutants have a different toxicity, atmospheric residence time and chemical stability than the original pollutant, with most of them being in the condensed phase. The formation and growth of secondary particles are an example of chemical conversion. Finally, deposition processes – dry or wet - are also produced, thus reducing the atmospheric levels. Dry deposition is considered to be the removal of particles and gases by sedimentation, diffusion or impaction processes and their subsequent physical/chemical attachment to terrestrial surfaces. On the other hand, wet deposition encompasses all processes by which pollutants are transported to the Earth's surface in aqueous form.

The concentrations and atmospheric residence times of primary or secondary exhaust pollutants depend on chemical and physical scavenging processes [5,6]. The time scales of these atmospheric transformations and loss flows vary widely. Reactive compounds, such as aromatic hydrocarbons, are removed from the atmosphere relatively quickly with atmospheric residence time of seconds. By contrast, less reactive pollutants, such as CO or CO_2, can be transported over greater distances and have an atmospheric residence time of months.

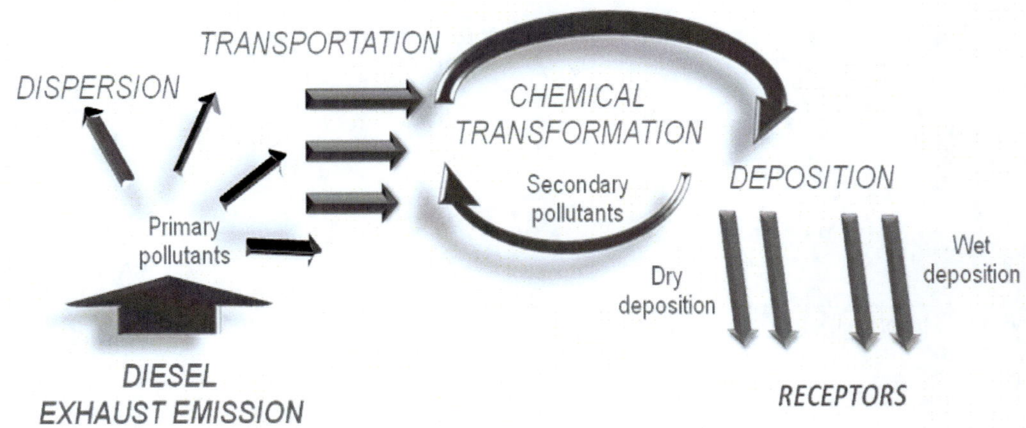

Figure 1. Atmospheric processes involving diesel exhaust emissions.

In response to requests by governments and industry, numerous scientific reports and assessments have been made on the potential impacts of current and future diesel engines. Although information on the total chemical and physical nature of diesel emissions is widely available, the knowledge required to accurately and fully characterize the gas and the particle phase of these emissions. It is known that hundreds of organic compounds are emitted from the complete and incomplete combustion of diesel fuel and that the composition depends on the type of engine, the type of fuel, the catalyst, and the engine operation cycles [7]. In addition, their chemical conversions depend on exhaust composition, concentration of oxidants, sunlight intensity and atmospheric conditions. Thus, the study of diesel emissions requires full-screening methodologies, suitable for monitoring both stable primary pollutants and products of chemical transformations. The main instrumental facilities used for monitoring direct or short-time exhaust emissions are dynamometer testing (chassis or engine), dilution tunnels, remote sensing and roadway assemblies. Nevertheless, simulator photoreactors – especially large-volume chambers - can also permit a representative monitoring of both directly emitted compounds and secondary species from different chemical transformations. The purpose of this chapter is to evaluate the potential of this analytical facility as a tool for understanding the chemical processes that take place after the diesel pollutant release. For this, our first goal is to compile and evaluate scientific information related to the monitoring of these primary and secondary exhaust emissions and to review the state of knowledge concerning the various aspects of this problem, such as diesel exhaust composition and atmospheric degradation processes that can affect it.

CHEMICAL TRANSFORMATIONS OF DIESEL EXHAUST EMISSIONS

A) Primary Diesel Emissions

The diesel exhaust engine is a complex mixture of hundreds of organic and inorganic constituents in the gas and particle phases, see Table 1.

Table 1. Chemical composition of primary diesel emissions, adapted from US EPA [8].

	Inorganic compounds	Organic compounds
Gas phase	CO_2, CO NOx (mostly NO+NO_2) SO_2 Ammonia	Alkanes and Cycloalkanes (<C20) Alkenes (C2–C4) Carboxylic acids (<C4) Monocyclic Aromatic Compounds PAH (<5-ring), Alkyl-PAH (<4-ring) Phenols and carbonyl Compounds
Particle phase	Elemental carbon Sulfates, nitrates Trace metals Water	Alkanes and Cycloalkanes (>C15) n-Alkanoic acids and diacids n-Alkenoic acids Aromatic acids and dicarboxylic acids PAH (>2-ring), Alkyl-PAH (>2-ring), Oxy-PAH, Nitro-PAH Heterocyclic compounds Hopanes/Steranes

The complete and incomplete combustion of fuel in a diesel engine results in the formation of gas phase components [1,5]. The most abundant of these are carbon oxides, nitrogen oxides - predominantly in the form of NO- and sulfur dioxide. Gas-phase organic compounds from diesel exhaust include a variety of paraffins, olefins, aromatics and their oxygenated analogs.

Diesel exhaust particles are aggregates of spherical primary particles consisting of solid carbonaceous material, with adsorbed organic compounds and small amounts of sulfate, nitrate, metals, trace elements, water, and unidentified compounds [5]. Organic material generated from unburned fuel, evaporated lube oil and pyrosynthesis, is mainly composed of alkanes, carboxylic acids, condensed polycyclic aromatic hydrocarbons (PAHs) and their oxidized products. Most diesel exhaust particle mass is associated with the accumulation mode particles, ranging from 0.05 to 0.7 mm, centered at about 0.2 mm. Thus, nearly all of the particle mass emitted is in the fine particle size range and nearly all the particle number emitted is in the nanoparticle diameter range [9-11].

The concentration of direct exhaust gases and particles depends on the amount of diesel fuel mass burned in the engine and the amount of air mass mixed with them [4,5]. Diesel exhaust composition also depends on several factors, i.e., fuel composition: diesel, reformulated diesel and source of biodiesel; engine type: light-duty (LD) and heavy-duty (HD); and engine operating conditions: driving cycles or individual regimes [1,12]. Additional factors are the presence of catalysts and working conditions such as temperature, pressure and humidity. In particular, the effects of diesel reformulation and engine operating parameters have an important impact on the gas-and-particle phase composition. For example, incomplete combustion in diesel engines running under low load, produces a relatively low particle concentration and a higher proportion of organic-associated particles. The result is that the percentage of elemental carbon in relation to total carbon varies from 70-90%, depending on vehicle type and operating conditions. The remainder corresponds to organic carbon, of which only about 20–30% have been characterized.

B) Atmospheric Transformations of Diesel Exhaust Emissions

To understand the dynamic composition of diesel exhaust emissions it is firstly necessary to determine the chemical changes produced by their reaction with atmospheric species. Tropospheric chemistry involves both the participation of reactive gaseous species such as radicals or ozone and the contribution of sunlight [13,14]. Reactions can occur in the gas phase or in the particle phase, and they follow several reaction pathways [4,15]. Moreover, the rates and mechanisms of reactions depend on the abundance of reactants, the composition of diesel exhaust and the temperature. For instance, the initial partitioning of the organic compounds between the gas and particle phases significantly influences the relative importance of the plausible degradation routes.

Figure 2 describes the main atmospheric reaction pathways followed by diesel exhaust constituents, including their relative importance and the contribution of sunlight. The most important degradation reaction of diesel exhaust is the daytime reaction with OH radicals. Other chemical processes for some diesel pollutants are photolysis, reaction with O_3, and reactions with NO_3 radicals during nighttime hours [16].

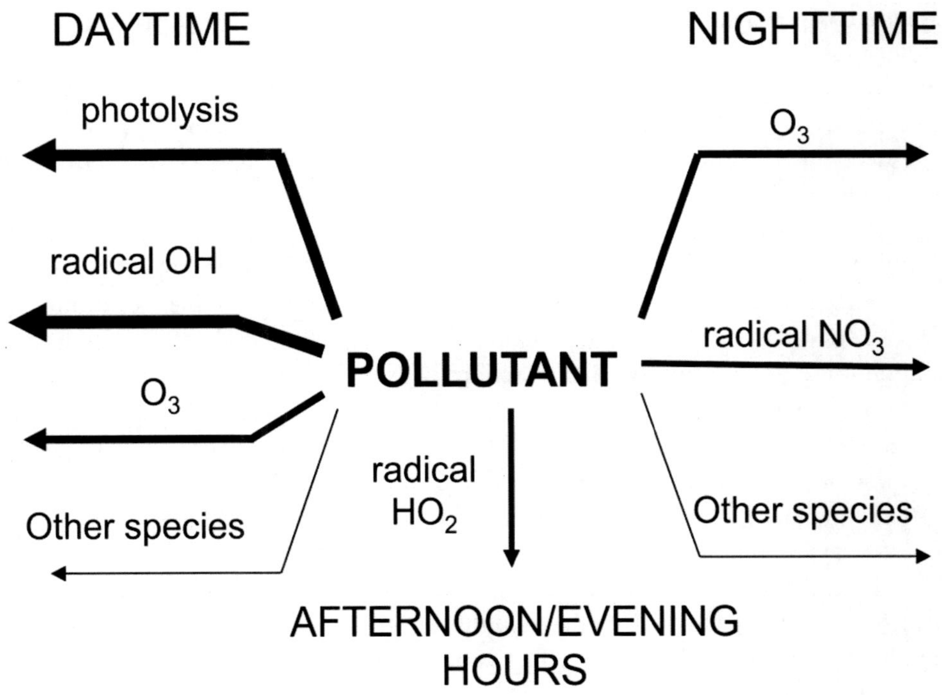

Figure 2. Atmospheric reactions followed by diesel emission constituents. The thickness of the line represents the relative importance.

The key parameter for describing the chemical removal of emissions is the atmospheric lifetime or time to decay to 1/e of its original concentration. The final atmospheric lifetime is a combination of different gas-phase reactions and it also depends on the ambient concentration of the tropospheric species involved. The lifetimes of automotive emissions are experimentally determined from the corresponding measured reaction rate constants of each individual process [15].

For inorganic gas phase components, several studies have investigated the chemical changes of diesel emissions suggesting the following routes and products [5]. In the early phase, the air chemistry is essentially unaffected by photolysis (the time scale of NO_2 photolysis is in the order of 100 s). The emitted NO reacts quickly with O_3 to produce NO_2 and O_2. At the same time, NO_2 can photolyze, react with O_2 and regenerate the original species, closing the cycle. During nighttime, essentially all NO gets oxidized this way. At the engine exit channel, the exhaust contains relatively large concentrations of hydroxyl and other radicals which oxidize parts of the emitted NO, NO_2, and SO_2 to nitrous acid (HONO), nitric acid (HNO_3), and sulphuric acid (H_2SO_4) by the formation of the HO_2 radical and HSO_3. The conversion depends strongly on the amount of OH and other radicals which are available at the engine exit and are consumed by these reactions. The conversion rates are important because the resultant gases, HNO_3, H_2SO_4, together with H_2O, are important precursor gases for particle formation and acid rain. The total increase in NO_2 enhances the photochemical generation of O_3 by photosmog reactions with CO and hydrocarbons.

The oxidizing environment of the atmosphere explains the transformations of the organic gas-phase components [4,17]. The dominant reaction pathway for alkanes is the reaction with OH radicals to form aldehydes and ketones, which reacts further under atmospheric conditions and may partition into the particle phase to form secondary organic aerosol (SOA). Products of alkenes after reaction with OH radicals, NO_3 radicals, and O_3 include aldehydes, ketones or both. The first generation of products from the degradation of monocyclic aromatic hydrocarbons include aromatic aldehydes, alkyl-benzyl nitrates, phenols, glyoxal, methylglyoxal and other saturated dicarbonyls, ring-opened unsaturated dicarbonyls, epoxides, quinines, nitro-PAHs, and hydroxy-nitro-PAHs. The major loss processes for these oxygenated products involves photolysis and reaction with OH radicals, forming products such as peroxyacetyl nitrate. For example, acetaldehyde forms peroxyacetyl nitrate (via formation of peroxyl radicals and reaction with NO_2) and benzaldehyde, the simplest aromatic aldehyde, forms peroxybenzoyl nitrate or nitrophenols following reaction with nitrogen oxides. Moreover, gas-phase PAHs react with the OH radical as their major removal route, although the contribution of their reaction with the NO_3 radical is also important.

In like manner the atmospheric chemistry of diesel exhaust impacts particulate matter in two basic ways: soot aging and secondary aerosol formation [18,19]. The elemental carbon in particulate diesel exhaust is inert to atmospheric degradation, whereas the organic fraction is degraded by heterogeneous reactions with different radicals and gaseous species. For example, particle-bound PAHs undergo photolysis, nitration, and oxidation reactions. In addition to changes in particle composition by degradation reactions, particle size distributions may vary depending on aggregation and coagulation phenomena in the aging process. Secondary particles may also form by nucleation of sulphuric acid with water on the surface of emitted soot particles or by photo-oxidation of organic compounds.

In conclusion, secondary pollutants from automotive emission constituents are mainly products of reaction with OH radicals, a degradation process with a high reaction rate. Nevertheless, other slower reactions also need to be studied because of the potential chemical or biological importance of their products. For example, the reaction of gas-phase PAH with NO_3 appears to be of minor significance but it is an important formation route for nitro-PAH, which are mutagenic and can have grave consequences on human health [20]. Furthermore, the rate constants and mechanisms of several reactions with ambient reactants are still uncertain –i.e., reaction products of the carboxylic acids with high molecular weight are currently unknown-. This absence of information is a consequence of the complexity of the chemical processes involved in exhaust emissions and the problems associated with having to carry out some laboratory studies, to work under unrealistic conditions.

MONITORING OF DIESEL EXHAUST EMISSIONS

In-line exhaust pipe systems are used to control the direct emission of automotive engines from cars and trucks. These measuring instruments from engine control and monitoring are used by well-known vehicle manufacturers, engine manufacturers, research institutions and government agencies. Historically, dynamometer testing is most widely used to estimate vehicle emission rates [21,22]. Primary diesel emissions are measured following well-established test procedures based on the driving cycles. For LD vehicles, pollutant levels are determined by chassis dynamometers and emissions are expressed in the mass distance scale (g/mile, g/km). HD diesel emissions are predominantly measured using engine dynamometers and they are expressed in mass/work units, mostly in grams/brake-horsepower hours (g/bhp-h) or sometimes in grams per kilowatt hour (g/kW-h).

For the dynamic evaluation of diesel emission composition, several strategies have been described. A simple approach is to use a combination of data on direct emissions and atmospheric conditions, as inputs in the dispersion equations [5,23]. The results of these air quality models are used to describe the emission composition as a function of time and to extrapolate the impact of diesel vehicles on atmospheric pollution. In recent years, the scientific effort has also been focused on the measurement of realistic vehicle emissions [24,25]. The approaches to this can be divided into in situ methods, in which monitoring equipment is deployed at fixed points (roadside or road tunnels), and in-traffic methods, in which monitoring equipment is deployed in probe vehicles driven within the local traffic stream. Both have provided valuable information; however these systems have the problem of lack of reproducible conditions and uncontrolled dispersion and interaction with the exhausts of other vehicles or other emission sources. In an intermediate approach, several studies have determined the size distribution of diesel aerosol in closed environments such as mines [26]. Nevertheless, for investigating causal relationships among variables, laboratory-based experiments are clearly the most appropriate strategy.

Full-equipped dilution tunnels are complex facilities to evaluate the composition of emissions under fixed conditions and real driving cycles [11,27,28]. As an example, Ford and GM manufacturers have constant volume sampling tunnels with several integrated monitoring systems. In these laboratories, the size distribution of soot particles has been studied routinely, accompanied by total hydrocarbon, CO, NOx, and CO_2 gaseous measurements. Their main limitation is that the dilution of diesel exhaust under roadway conditions is not the same as that produced in many of the dilution tunnels used in dynamometer studies. This change in dispersion volume, temperature, and humidity can potentially lead to significant differences in the nature of the emissions, especially the composition of particulate matter. Thus, the discrepancy can leads to laboratory-based predictions on particle size distribution and gas/particle phase partitioning of semi-volatile compounds, which differ slightly from those found under real driving conditions. Moreover, the relevance of these measurements for describing the final contribution of diesel exhaust to the atmosphere is always a question, due to the limited time it interacts with other ambient air species; even a dilution step is included.

An experimental approach to studying the atmospheric transformations of diesel exhaust emissions, involves the use of small reactors. These consist of reaction chambers with a light source to generate radicals, analytic devices, and a gas line for introducing the diesel exhaust and reactants into the cell. The Pyrex chamber at Ford Motor Company in Michigan (USA) is an example of the small evacuable chambers (0.14 m^3) used in exhaust studies [29]. The Toyota Company has used a 2 m^3 chamber facility in Aichi (Japan) and they have sponsored the construction of a chamber of the same size at Tsinghua University in China [30].

However, for representative results, a much larger reaction vessel is required.. Large chambers minimize chamber surface area-to-volume ratios, thereby reducing wall effects such as the adsorption and desorption of gas-phase species or particle compounds. In this sense, the photochemical reactors or 'smog chambers' are invaluable tools in the investigation of atmospheric sciences [31-33]. The simulators allow the examination of specific atmospheric chemical transformations, since precise oxidant conditions can be reproduced. The photochemical reactor applications include determinations of lifetimes and relative rates of reaction, studies on product distributions and development of degradation mechanisms. An example of a full-equipped photo-simulation chamber is the European Photoreactor (EUPHORE). This facility, with two photoreactors (200 m^3), is currently one of the largest and best-equipped outdoor simulation chambers in the world.

SMOG CHAMBER FOR DIESEL EXHAUST MONITORING

A) Instrumentation

The basic elements of a photoreactor are the chamber structure, the control system, the light source, the injection/formation modules of reactants, the exhaust emission injection system, the mixing components and the monitoring instruments for precursors and products [30,34]. With respect to the light source, we have to consider that the atmospheric reactions are induced if adequate concentrations of radicals and molecules -present under daytime or nighttime conditions- are achieved [15]. Thus, photoreactor chambers can be divided into two categories from the point of view of light generation: those driven by natural sunlight and those driven by artificial light [31]. The first type of photoreactors consists of outdoor structures with transparent material (e.g., FEP fluorine-ethene-propene). Because this foil allows more than 80% light transmission in the UV visible region (280 – 640 nm) solar radiation pass through it. The use of these chambers provides less reproducible ambient conditions due to day-to-day variation. The use of natural sunlight in the chambers induces the photolysis reactions properly (broad wavelength spectra). This variability can however be minimized by correcting the results with the radiation parameters JNO_2 (light intensity) and θ (zenith angle).

Large photoreactors are equipped with a control system designed to select and record functional parameters (valves for injection, mixing devices, etc.). Several measurements systems are also integrated for measuring physical parameters (radiation, pressure, humidity and temperature) and monitoring chemical parameters such as precursor species, intermediate and final products [33-36]. The JNO_2 value of light is measured with a spectral radiometer. Pressure, humidity and temperature are measured by means of specific sensors. Most photoreactors are equipped with a white-type mirror system coupled to a Fourier Transform Infrared spectrometer with a long absorption path length, and a high spectral and time resolution. Simultaneously, different monitors of the main atmospheric gases are integrated for measuring in the low concentration ranges (NO_x, O_3, SO_2, CO, CH_4, $HCHO$, $HONO$, etc). The laser-induced fluorescence instrument (LIF) allows the monitoring of OH and HO_2 radicals. Full-equipped smog chambers have several on-line analytical instruments for measuring gaseous compounds: gas and liquid chromatographs with several detectors, differential optical absorption spectrometer (DOAS) and proton transfer resolution – mass spectrometer (PTR-MS). For diesel aerosol size characterization two types of instruments are used: those that provide size distribution data and those that collect samples for subsequent analyze such as gravimetric, chemical or microscopic analysis [9]. Aerosol mass spectrometers (AMS), nephelometers, scanning mobility particle sizers (SMPS), tapered element oscillating monitors (TEOM) and differential mobility analyzers (DMA) are typically installed in simulator chambers. These instruments provide information on physical properties such as size distributions, aerosol yielding and concentration profiles. In addition, the integration of cartridges or filter systems allows a selective sampling to be made of specific families of gaseous or condensed compounds for subsequent analysis.

B) Smog Chamber Experiment for Exhaust Monitoring

A diesel exhaust experiment in a photoreactor consists of studying a specifically induced reaction of pollutants with atmospherically relevant radicals under controlled conditions, isolated from meteorological variations or dispersion [32]. For this, the photoreactor is filled with air in the adequate composition - fixed concentrations of reactants- and mixed with diesel exhaust under a time-controlled light exposition. Generally speaking, the steps in a large photoreactor experiment are: set-up of chamber facility, background determinations, reaction monitoring and chamber cleaning [34].

Several maintenance and set-up activities must be performed. The calibration of the analytical instrumentation is achieved by injecting standard compounds into the photoreactor. Checking for leaks and the adjusting the temperature and relative humidity are necessary to avoid experimental variations. Afterwards, a cleaning period is carried out by flushing the chamber with air –either purified or synthetic air-, to assure that there are no residual amounts of reagents from previous experiments is in the reactor. The use of purified air, obtained by filtration and drying, requires a cleaning time around of 12h, but small trace levels of CO and CH_4 can not be totally eliminated. Synthetic air is more expensive but leaves no trace reactants in the photoreactor. In this case, chamber is flushed by dilution and more than three days are needed to be sure that no reactants are left in the chamber. An over-pressure is maintained in the chamber, to avoid contamination from outside components. Then, on-line instruments measure the background levels of the analytical parameters, checking that the initial conditions are within the operational limits. Finally, a tracer compound is injected into the chamber to quantify the dilution factor - leaks and sampling systems -. The dilution factor is used for correcting all the gas and particulate data recorded during the experiment. SF_6 is commonly used, since this gaseous compound is inert, stable at ambient temperatures to all radical processes and light and it features, easy-injection, low cost and easy monitoring by FT-IR.

The next step consists in introducing the reactants into the chamber through a system of valves and Teflon transfer lines. If the reactant interacts with Teflon, as in the case of amines and their oxygenated products, sulfiner or stainless steel lines must be used [34]. Different injection strategies are used according to the nature and reactivity of the reactants. Gas compounds, such as NO, NO_2, SO_2, are directly provided from the calibration bottles. Semi-volatile liquids or semi-volatile solids are vaporized in an impinger and carried by heated air stream. For high amounts of liquids –i.e., water or hydrogen peroxide-, small drops are generated by a sprayer system. Compounds with high reactivity, such HONO or O_3, require being freshly synthesized. A conventional method to form HONO is the liquid-phase reaction of $NaNO_2$ and H_2SO_4. A simple method to generate ozone in situ is to pass a flow of pure oxygen through a lamp that emits at 254 nm, thereby transforming O_2 in O_3. In some experiments, seed aerosols are also introduced into the chamber to promote aerosol formation – nucleation seeds - and to force some degradation routes – solid phase reactions -. Seeds are introduced from a concentrated solution of NaCl, Na_2SO_4 or KNO_3 by an aerosol atomizer.

In the case of diesel experiments, the exhaust introduction protocol is critical to minimize composition changes due to water condensations, dilution effects or interactions with the injection system materials [8,37,38]. Diesel exhaust is generated on-site, using diesel engines assembled in dynamometers [16,32,39]. The diesel engines can be operated under different driving conditions or using fuels of different composition. A heated Teflon transfer line is recommended for connecting the exhaust pipe to the chamber. The system can be equipped with a gas analyzer for the continuous monitoring of engine-out gaseous emissions. A dilution system can also be included to reduce the initial concentration of pollutants. Alternatively, a filter system can be inserted to eliminate soot particles, totally or partially according to particle size. The injection order of reactants/diesel depends on the type of degradation reaction. For example, in ozonolysis experiments, the diesel exhaust is introduced first, due to the fact that hydrocarbons react quickly with ozone.

Mixing fans perform a rapid homogenization of the exhaust components with the reactants. This prevents air stratification, agglomeration of particles and the water condensation and it assures a representative sampling in the reaction chamber.

Table 2. Smog chamber experiments for studying the chemical transformation of diesel exhaust emissions

Objective	Oxidant reagent	Light exposure
Diesel exhaust characterization	-	No
Ozonolysis reaction	Ozone	No
Nitrate radical reaction	NO_3 radical	No
Photolysis	-	Yes
Photo-oxidation in the presence of NOx	HONO	Yes
Photo-oxidation in the absence of NOx	H_2O_2	Yes

For nighttime experiments (oxidations), the start of the reaction is considered to occur when all components are mixed. For daytime reaction experiments (photolysis and photooxidations), the chamber is generally opened (outdoor chambers) or the lights are switched on (indoor chambers) after mixing. Table 2 shows a summary of the main types of experiments.

The experiment duration ranges between 0.5-48 h, according to the following criteria: the aerosol level or the formation of a selected product is reached, all the reactants are consumed or the light exposition time is exceeded. When the experiment has finished, the flushing system is connected and injects clean air into the chamber until the background levels are reached. In the case of diesel exhaust experiments, a special cleaning process must be performed since carbonaceous soot is retained and deposited on the foil cover of the chamber, thus interfering with later studies.

C) Limitations of Smog Chambers

At present, and despite their scientific potential, the most important limitation to using smog chambers for diesel exhaust studies has to do with demonstrating the validity of their results [31,32]. Previous characterization of the smog chamber is essential, since some fundamental parameters need to be obtained. The exact loss rates of gaseous reactants and particles should be calculated from profile on trace compounds or control experiments. The absence of residual concentrations of pollutants or reactants should be demonstrated to avoid undesired reaction pathways [15]. Also, for demonstrating the reproducibility, experiments must be conducted several times under the same conditions i.e., engine operating mode, diesel exhaust time injection, oxidant levels, light exposure time, radiation intensity, etc.

The study of a specific reaction is associated with a set of experiments. For example, the evaluation of diesel exhaust photo-oxidation degradation in the presence of NOx should require the following assays. First, calibration and background experiments will set-up and characterize the operational conditions and the analytical instrumentation of the photoreactors. A light exposed experiment, where only diesel exhaust is injected into the chamber, will be performed to characterize its chemical composition and photolysis. In dark conditions, the diesel exhaust will be mixed with NOx to monitor plausible oxidations in the absence of light. Finally, the photo-oxidation degradation will be simulated after diesel exhaust-NOx mixing under light exposure. Moreover, the products formed from diesel exhaust degradation change depending on the concentration of reactants, experiments with different ratios diesel/reactant should also be performed. In addition, the specific drawbacks of outdoor chambers are that the experiments are limited to days with appropriate meteorology and temperature control becomes difficult [15]. Finally, a full screening of gaseous and condensed compounds involves a high number of instruments simultaneously integrated. Because of the low concentration and high instability of several critical species, the analytical requirements for monitoring systems are important.

DIESEL EXHAUST STUDIES IN SMOG CHAMBERS

Atmospheric studies of interest for compressive analysis of exhaust pollutants have been performed in different smog chambers, listed in Table 3. For instance, the reactivity of some volatile organic compounds or particles has been evaluated under different atmospheric conditions [29,31,40]. However, the number of papers based on experiments with freshly emitted diesel exhaust is still reduced. Tedlar bags were used to collect the exhaust sample prior to its transfer to General Motors smog chamber [41]. The chambers run by the University of North Carolina (UNC), Carnegie Mellon University (CMU) and California Institute of Technology (CALTECH) have facilitated the investigation of atmospheric aerosol chemistry. They have studied the formation and aging of SOA from the photochemical oxidation of diesel exhaust under different conditions. Diluted diesel soot was exposed to sunlight and high ozone concentration to evaluate aging [42]. The impacts of diesel soot on SOA formation has been demonstrated in presence of a biogenic compound and upon exposure to atmospheric oxidants [43,44]. The smog chamber data provided strong evidence that the oxidation of diesel exhaust compounds, which are not accounted for in existing models, contributes to ambient SOA formation [45]. On other hand, photo-oxidation of markers associated with unburned lubricating oil for were induced after flash vaporizing motor oil [46]. Recently, the oxidation kinetics of these markers combined to motor vehicle exhausts was also examined [47].

At the large aerosol chamber facility AIDA (Aerosols, Interactions and Dynamics in the Atmosphere), an intensive campaign was organized [48]. A comprehensive physical and chemical characterization of soot aerosol was performed with the participation of several research groups.

Table 3. Smog chambers used for studying pollutants related to diesel exhaust emissions

Chamber	Location	Volume	Webpage
GM R&D center	Warren, MI, USA	$0.7\ m^3$	http://www.gm.com/
Ford	Dearborn, MI, USA	$0.14\ m^3$	http://www.ford.com/
Toyota R&D Labs	Aichi, Japan	$2\ m^3$	http://www.tytlabs.co.jp/
UNC chambers	Pittsboro, NC, USA	$270\ m^3$	http://www.sph.unc.edu/envr
CMU chamber	Pittsburgh, PA, USA	$10\ m^3$	http://caps.web.cmu.edu/
CALTECH	Pasadena, CA, USA	$28\ m^3$	http://www.che.caltech.edu/
CE-CERT	Riverside, CA, USA	$90\ m^3$	http://www.cert.ucr.edu/
EUPHORE	Valencia, Spain	$200\ m^3$	http://euphore.es/
SAPHIR-AEC	Jülich, Germany	$270\text{-}280\ m^3$	http://saphir.fz-juelich.de
AIDA	Karlsruhe, Germany	$84\ m^3$	http://imk-aida.fzk.de
PSI	Villigen, Switzerland	$27\ m^3$	http://lac.web.psi.ch

Recent projects have been designed to study diesel emissions in EUPHORE, a partner of EUROCHAMP (Integration of European Simulation Chambers for Investigating Atmospheric Processes). Figure 3 shows a diagram of the instrumentation used at EUPHORE for diesel experiments. The EC project DIFUSO (Diesel fuel and soot: fuel formulation and its atmospheric applications) was focused on the investigation of the impact of real diesel exhaust on atmospheric chemistry [39,49]. Mixtures of volatiles organic compounds plus nitrogen oxides were irradiated by natural sunlight in the presence and the absence of diesel exhaust. An important result was to demonstrate that the increase in ozone formation observed on adding diesel exhaust was mainly caused by the exhaust concentrations of nitrous acid and formaldehyde (radical sources). To elucidate the potential toxic effects of diesel exhaust, secondary compounds formed under different degradation reactions have also been studied [32]. In the following section, we describe the characterization of condensed

PAHs from diesel exhaust emissions. Some results have recently been published in Borrás et al. 2009 [50].

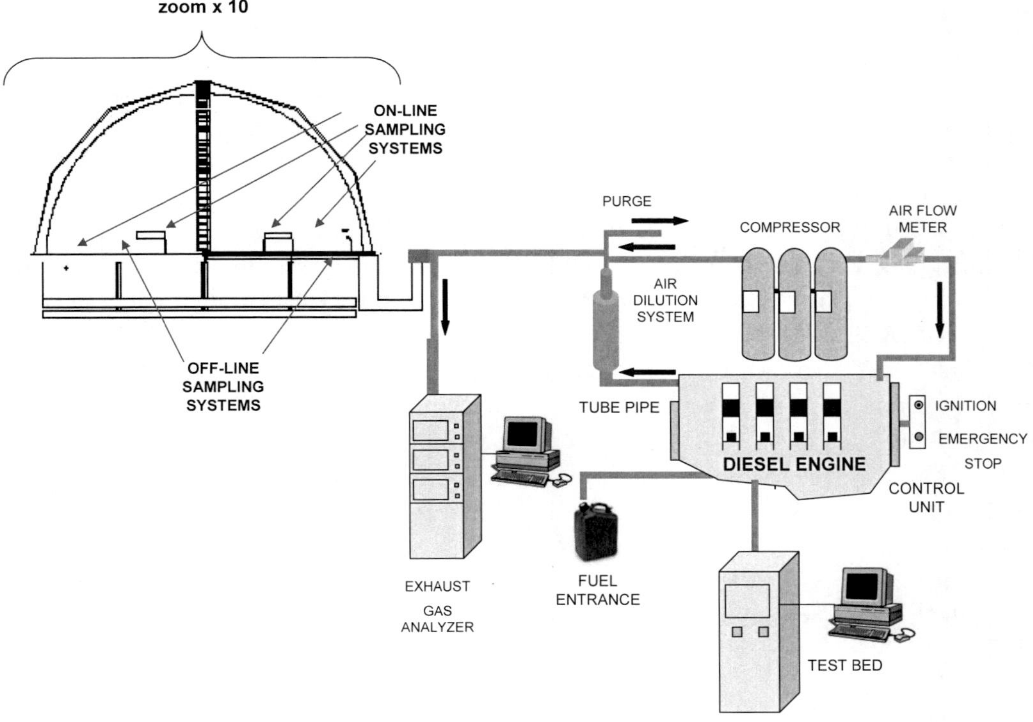

Figure 3. Diagram of the instrumentation used in EUPHORE experiments.

CASE STUDY: MONITORING OF PAHS

Diesel vehicles produce high particulate emissions, and PAHs are considered to be the most relevant organic compounds produced [18,51,52]. Low molecular weight (LMW) PAHs in an unburned diesel fuel have been shown to be the primary contributor to the exhaust emissions from a direct-injection diesel engine. Medium molecular weight (MMW) PAHs, such as fluoranthene and pyrene, and high molecular weight (HMW) PAHs, such as benzo[b]fluoranthene and benzo[a]pyrene, are formed during the combustion processes. Total PAHs generally constitute less than 1% of the diesel particulate matter mass. However, because degraded PAH products are often more mutagenic than their precursors, the formation, transport, and concentrations of these compounds in diesel aerosols are of significant interest [53-55].

The reformulation of diesel fuels, e.g., by the modification of aromatic compounds or sulfur content and/or by developing alternative fuels [38], could achieve substantial reductions in PAH exhaust emissions. Exhaust emissions from diesel vehicles depend on engine type: light-duty or heavy-duty engine [56] and engine operating conditions: driving cycles or individual regimes [12]; however, most of the literature deals with complete driving cycles. In conclusion, a better description both of PAH exhaust emissions under different conditions and of their relationship with gaseous pollutants is still needed in order to determine the best strategies to reduce them.

EUPHORE photoreactors have been used to evaluate exhaust emissions and their chemical degradation processes. The aim was the monitoring and systematic investigation of PAH compounds emissions to observe and quantify differences between diesel compositions, engine driving cycles, and chemical degradations. For this, an engine test bed (brake dynamometer) was used to control engine operation conditions, registering parameters such as speed, torque and fuel consumption. Also, a gas analyzer was used to directly measure gaseous exhaust concentrations while a filter sampling system collected direct particulate matter. The rest of the emission was diluted and guided through heated tubes to the simulation chamber. The chamber experiments were designed to characterize the chemical composition of the diesel exhaust – gaseous and particulate -, to examine the effects of aging, photolysis and reactions with several radicals on the diesel exhaust composition.

First, short-time diesel exhaust emissions were evaluated to assess the PAH particulate concentrations as a function of fuel chemical composition. Fuels used were: standard diesel fuel, reformulated diesel fuels with different percentages of hydrocarbons, diesel fuels with different sulfur contents and biodiesels with different origins (palm, rape oil etc.). The main PAH compounds observed – EPA's priority chemical list and several isomers- are shown in Table 4.

Table 4. PAH compounds detected in EUPHORE experiments (oxy-PAHs and nitro-PAHs are not included)

EPA PAHs		Other PAHs
Naphthalene	Benzo[a]anthracene	Methylnaphthalene isomers
Acenaphthylene	Chrysene	Dimethylnaphthalene isomers
Acenaphthene	Benzo[b]fluoranthene	Trimethylnaphthalene isomers
Fluoranthene	Benzo[k]fluoranthene	Methylfluorene isomers
Phenanthrene	Benzo[a]pyrene	Methylphenanthrene isomers
Anthracene	Indene[1,2,3-cd]pyrene	Methylanthracene isomers
Fluoranthene	Benzo[g,h,i]perylene	Dimethylphenanthrene isomers
Pyrene	Dibenzo[a,h]anthracene	Dimethylanthracene isomers
		Benzoanthracene isomers
		Benzopyrene isomers
		Benzofluoranthene isomers

The results showed that the PAH concentrations from reformulated diesel fuels with different percentages of aromatic hydrocarbons were quite similar (3.6 ± 0.6 ngPAH mg^{-1}). Thus, a variation of the diesel fuel aromatic content did not significantly reduce PAH emission. For the biodiesel exhaust emissions, the 16 EPA PAH compounds were identified; four of them, chrysene, benzo[b]fluoranthene, benzo[k]fluoranthene, indene[1,2,3-cd]pyrene were present only in these biodiesel exhaust emissions. Moreover, although the exhaust emissions from biodiesel were characterized by a greater number of compounds, the use of biodiesel reduced the total concentration of PAHs by around 50 % with respect to the reformulated diesels.

The PAH concentration from diesel exhausts emitted by different engine driving cycles were individually evaluated, and some differences were found between them. The cold start values ranged from 0.05 - 9 µg g^{-1} for LMW, 5 - 20 µg g^{-1} for MMW and 1 - 150 µg g^{-1} for HMW PAH compounds. Meanwhile, the idling values were 0.3 – 27 µg g^{-1} for LMW, 2 – 40 µg g^{-1} for MMW and 6 – 70 µg g^{-1} for HMW. During the cold start cycle (higher power mode), the concentration of all PAHs was high, especially the HMW compounds like chrysene, benzo[k]fluoranthene, benzo[a]pyrene and dibenzo[a,h]anthracene, confirming that cold start operating conditions generate high concentrations of HMW PAHs. In maximum fuel consumption conditions of idling, PAH emissions also became associated with HMW compounds. An analysis of PAH exhaust emissions was also performed to compare accelerating and steady-state engine conditions The total PAH exhaust emissions produced during acceleration were on average 1.5 times greater than the emissions produced by the steady-state conditions. Moreover, during acceleration cycles, concentrations of HMW PAHs showed an increase of more than twice compared to LMW PAHs. In steady-state, differences in exhaust concentrations have been observed for total LMW PAHs (0.1 – 7 µg g^{-1}) and HMW PAHs (3 – 53 µg g^{-1}). Statistical analyses showed trends between total PAH, CO_2, CO, total HC and NOx, depending on whether the combustion process was considered complete or incomplete. In summary, this study demonstrated that the driving pattern has a significant effect on diesel exhaust emissions from light-duty diesel engines.

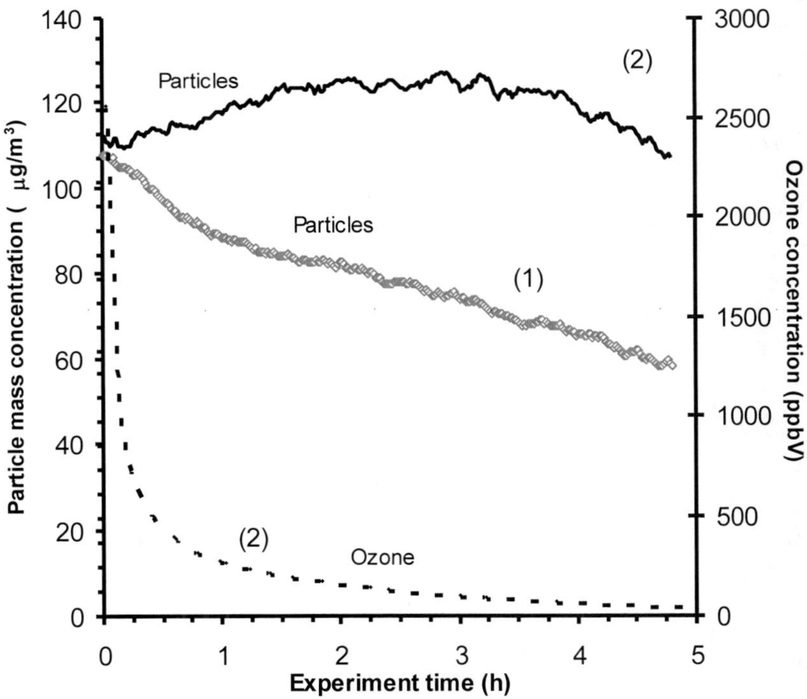

Figure 4. Aerosol profiles for ozonolysis experiment: (1) Only diesel exhaust; (2) Ozone and diesel exhaust.

Nevertheless, the main applicability of this type of large volume photoreactors is the detailed study of diesel exhaust degradation processes. As described above, different reactions were studied in the EUPHORE photoreactors under controlled and near realistic conditions. The experiments induced reactions such as photolysis, photo-oxidation in the presence or in the absence of NOx, and dark reactions with O_3 and NO_3 radical. Long-time diesel exhaust emissions were analyzed by different on-line and off-line systems. The mass concentration and size distribution of aerosols were continuously registered. Below, we show an example of TEOM profiles for diesel exhaust particulate matter from the EUPHORE photoreactor under different conditions (Figure 4).

Moreover, for determining the chemical particulate composition, quartz fiber filters were use to collect condensed PAH compounds in each condition. The filters were then extracted and analyzed by high resolution capillary gas chromatography-mass spectrometry (GC/MS). Figure 5 shows a typical gas chromatogram of PAH compounds from diesel exhaust photo-oxidation in the presence of NOx. Preliminary results show differences in the profile of PAHs and their oxidized products (oxy-PAHs and nitro-PAHs) as a function of time.

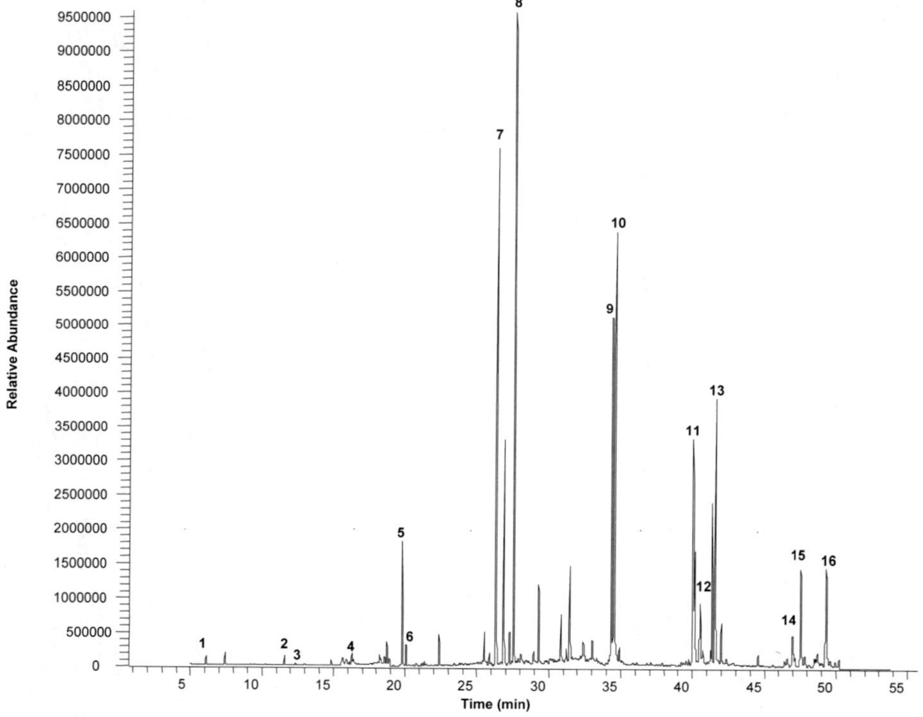

Figure 5. Chromatogram of diesel exhaust particulate extract of 16 EPA PAHs in SIM mode. 1. naphthalene, 2. acenaphthylene, 3. acenaphthene, 4. fluorene, 5. phenanthrene, 6. anthracene, 7. fluorantene, 8. pyrene, 9. benzo[a]anthracene, 10. crysene, 11. benzo[b]fluoranthene, 12. benzo[k]fluoranthene, 13. benzo[a]pyrene, 14. indene[1,2,3-c,d]pyrene, 15. benzo[g,h,i]perylene, 16. dibenzo[a,h]anthracene.

The research objectives are currently focused on searching for correlations between PAH profiles and the initial oxidant conditions (reactant concentration and light exposure), reaction yields and/or the chemical composition of gaseous and particulate phase. To better understand the chemical degradation pathways, several gaseous compounds -CO, CO_2, NOx and most volatile organic compounds- were also monitored. Statistical tools will be applied to facilitate the complete analysis of the data. The expected results are that the concentrations of oxidized PAH products, including nitro-PAHs, will be found to significantly increase under aggressive atmospheric conditions. The results will also involve a direct tracking of these human toxic compounds derived from diesel exhaust emissions.

This kind of information is highly useful for developing technological solutions in the automotive industry and for planning quality control strategies. As a final conclusion, it is important to point out that fully-equipped photoreactors are the only scientific facilities capable of providing data on the realistic influence of atmospheric chemical conditions in the composition of diesel exhaust emissions, avoiding the interference of other pollution sources.

CONCLUSION

The impact evaluation of diesel exhaust emissions on atmosphere requires characterizing both primary and secondary compounds. For this, the composition of initial exhaust emissions must be determined, as well as a complete monitoring of products from their atmospheric transformations. As it has been demonstrated, large photoreactors are the best scientific facilities to provide data about the realistic influence of atmospheric chemical conditions in the degradation of diesel exhaust emission, avoiding the interference of other pollution sources or other physical processes. The chapter has shown the basic protocol of a photoreactor experiment and the instrumentation required to obtain accurate atmospheric data. So, different research groups have provided valuable results of both gas and particulate phase –i.e. the formation and aging of secondary aerosols, relationship between diesel emission and photosmog effect-. This kind of information is highly useful for developing technological solutions in the automotive industry. The progress of quality control strategies of exhaust emissions and the future selective regulations is also linked with the results of these atmospheric investigations.

ACKNOWLEDGMENTS

We gratefully acknowledge the Generalitat Valenciana, the Fundación Bancaja and the GRACCIE CBS2007-00067 project on CONSOLIDER-INGENIO 2010 program for supporting this chapter.

REFERENCES

[1] Heywood, J.B. Motor Vehicle Emissions Control: Past Achievements, Future Prospects. In: E. Sher, editor. *Handbook of Air Pollution from Internal Combustion Engines.* San Diego, CA: Academic Press. 1998. Pages 3-23.

[2] Twigg, M.V. Controlling automotive exhaust emissions: successes and underlying science. *Phil. Trans. R. Soc. A.* 2005, 363, 1013-1033.

[3] Seinfeld, J.H., & Pandis, S.N. (2006). *Atmospheric chemistry and physics: from air pollution to climate change* (2nd edition). New York: Wiley.

[4] Harrison, R.M. (2001). *Pollution: Causes, Effects and Control* (4th edition) Royal Society of Chemistry. The University of Birmingham, UK.

[5] Watson, A.Y., Bates, R.R., Kennedy, D.(1988). *Air pollution, the automobile and public health.* Washington, DC: National Academy of Sciences Press.

[6] Holton, J.R., J.A. Curry; J.A. (2003) *Pyle Encyclopedia of Atmospheric Sciences* (2nd edition). San Diego CA; Academic Press.

[7] York, A.P.E., & Tsolakis A. Cleaner Vehicle Emissions. *Encyclopedia of Materials: Science and Technology,* 2008, Pages 1-7

[8] US EPA Health assessment document for diesel engine exhaust, EPA/600/8-90/057F. 2002. Available from URL: *http://www.me.umn.edu/centers/cdr/Proj_EPA.html.*

[9] Kittelson, D. B. Engines and nanoparticles: a review. *J. Aerosol Sci.* 1998, 29, 575–88.

[10] Kleeman, M.J.; Schauer, J.J.; Cass, G.R. Size and composition distribution of fine particulate matter emitted from motor vehicles. *Environ. Sci. Technol.* 2000, 34, 1132-1142

[11] Burtscher, H. Physical characterization of particulate emissions from diesel engines: a review. *J. Aerosol Sci.* 2005, 36, 896–932

[12] Karavalakis, G.; Stournas, S.; Bakeas, E. Effects of diesel/biodiesel blends on regulated and unregulated pollutants from a passenger vehicle operated over the European and the Athens driving cycles. *Atmos. Environ.* 2009, 43, 1745-1752.

[13] Warneck, P. *Encyclopedia of Physical Science and Technology.* Tropospheric Chemistry, San Diego, CA: Academic Press. 2004. Pages 153-174

[14] Harrison, R.M. (2007). *Principles of Environmental Chemistry.* Chapter 2 Chemistry of the Atmosphere (2nd edition). Royal Society of Chemistry. The University of Birmingham, UK.

[15] Finlayson-Pitts, B.J. & Pitts, J.N. (2000) *Chemistry of upper and lower atmosphere* (1st Edition). San Diego, CA. Academic Press.

[16] Zielinska, B. Atmospheric transformation of diesel emissions. *Exp. Toxicol. Pathol.* 2005, 57, 31–42

[17] Atkinson, R. & Arey, J. Atmospheric Degradation of Volatile Organic Compounds. *Chem. Rev.* 2003, 103, 4605-4638.

[18] Maricq, M.M. Chemical characterization of particulate emissions from diesel engines: A review. *J. Aerosol Sci.* 2007, 38, 1079 – 1118.

[19] Kroll, J.H. & Seinfeld, J.H. Chemistry of secondary organic aerosol: Formation and evolution of low-volatility organics in the atmosphere. *Atmos. Environ.* 2008, 42, 3593–3624.

[20] Srogi, K. Monitoring of environmental exposure to polycyclic aromatic hydrocarbons: a review. *Environ. Chem. Lett.* 2007, 5, 169-195.

[21] Graham, L. Chemical characterization of emissions from advanced technology light-duty vehicles. *Atmos. Environ.* 2005, 39, 2385-2398.

[22] Andre, M. & Rapone, M. Analysis and modeling of the pollutant emissions from European cars regarding the driving characteristics and test cycles. *Atmos. Environ.* 2009, 43, 986-995.

[23] Wang, J. S.; Chan, T.L.; Cheung, C.S.; Leung, C.W.; Hung, W.T. Three-dimensional pollutant concentration dispersion of a vehicular exhaust plume in the real atmosphere. *Atmos. Environ.* 2006, 40, 484-497.

[24] Ropkins, K.; Beebe, J.; Li, H.; Daham, B.; Tate, J.; Bell, M.; Andrews, G. Real-World Vehicle Exhaust Emissions Monitoring: Review and Critical Discussion. *Crit. Rev. Env. Sci. Tec.* 2009, 39, 79-152.

[25] Kuykendall, J.R.; Shaw, S.L.; Paustenbach, D.; Fehling, K.; Kacew, S.; Kabay, V. Chemicals present in automobile traffic tunnels and the possible community health hazards: A review of the literature. *Inhal. Toxicol.* 2009, 21, 747-792.

[26] Bugarski, A.D.; Schnakenberg, G.H.; Hummer, J.A.; Cauda, E.; Janisko, S.J.; Patts, L.D. Effects of Diesel Exhaust Aftertreatment Devices on Concentrations and Size Distribution of Aerosols in Underground Mine Air. *Environ. Sci. Technol.* 2009, 43, 6737-6743

[27] Maricq, M.M.; Podsiadlik, D.H.; Chase, R.E. Examination of the size-resolved and transient nature of motor vehicle particle emissions. *Environ. Sci. Technol.* 1999, 33, 1618-1626.

[28] Isella, L.; Giechaskiel, B.; Drossinos, Y. Diesel-exhaust aerosol dynamics from the tailpipe to the dilution tunnel. *J. Aerosol Sci.* 2008, 39, 737 – 758.

[29] Hurley, M.D.; Sokolov, O.; Wallington, T.J.; Takekawa, H.; Karasawa, M.; Klotz, B.; Barnes, I.; Becker, K.H. Organic aerosol formation during the atmospheric degradation of toluene. *Environ. Sci. Technol.* 2001, 35, 1358-1366.

[30] Wu, S.; Lu, Z.F.; Hao, J.M.; Zhao, Z.; Li, J.; Takekawa, H.; Minoura, H.; Yasuda, A. Construction and characterization of an atmospheric simulation smog chamber. *Adv. Atmos. Sci.* 2007, 24, 250-258

[31] Cocker, D.R.; Flagan, R.C.; Seinfeld, J.H. State-of-the-art chamber facility for studying atmospheric aerosol chemistry. *Environ. Sci. Technol.* 2001, 35, 2594–2601.

[32] Barnes, I., & Rudzinski, K.J. (2006) *Environmental simulation chambers: application to atmospheric chemical processes.* NATO Science Series Dordrecht, The Netherlands. Springer.

[33] Nilsson, E. J.K.; Eskebjerg, C.; Johnson, M.S. A photochemical reactor for studies of atmospheric chemistry. *Atmos. Environ.* 2009, 43, 3029-3033.

[34] Becker, K.H. 1996. *Design and Technical Development of the European Photoreactor and First Experimental Results,* Final Report of the EC-Project (EV5V-CT92-0059), Wuppertal.

[35] Doussin, J.F.; Ritz, D.; Jolibois, R.D.; Monod, A.; Carlier, P. Design of an environmental chamber for the study of atmospheric chemistry: New developments in the analytical device. *Analusis.* 1997, 25, 236-242.

[36] Glowacki, D.R.; Goddard, A.; Hemavibool, K.; Malkin, T.L.; Commane, R.; Anderson, F.; Bloss, W.J.; Heard, D.E.; Ingham, T.; Pilling, M.J.; Seakins, P.W. Design of and initial results from a Highly Instrumented Reactor for Atmospheric Chemistry (HIRAC). *Atmos. Chem. Phys.* 2007, 7, 5371-5390.

[37] Moosmuller, H.; Arnott, W.P.; Rogers, C.F.; Bowen, J.L.; Gillies, J.A.; Pierson, W.R.; Collins, J.F.; Durbin, T.D.; Norbeck, J.M. Time resolved characterization of diesel particulate emissions. 1. Instruments for particle mass measurements. *Environ. Sci. Technol.* 2001, 35, 781-787.

[38] Kittelson, D.B.; Watts, W.F.; Johnson, J.P. On-road and laboratory evaluation of combustion aerosols—Part 1: Summary of diesel engine results. *J. Aerosol Sci.* 2006, 37, 913-930.

[39] Wiesen, P. (2000). *Diesel fuel and soot: fuel formulation and its atmospheric applications (DIFUSO)*. Final Report of the EC Project, Contract No. EV4V-CT97-0390, Wuppertal, FRG.

[40] Chang, T.Y.; Nance, B.I.; Kelly, N.A. Modeling smog chamber measurements of incremental reactivities of volatile organic compounds. *Atmos. Environ,* 1999, 33, 4695-4708.

[41] Chang, T.Y.; Nance, B.I.; Kelly, N. A. Modeling smog chamber measurements of vehicle exhaust reactivities. *J. Air Waste Manage.* 1999, 49, 57-63.

[42] Vartiainen, M.; McDow, S.R.; Kamens, R.M. Water uptake by sunlight and ozone exposed diesel exhaust particles. *Chemosphere.* 1996, 32, 1319-1325.

[43] Lee, S.D.; Jang, M.S.; Kamens, R.M. SOA formation from the photooxidation of alpha-pinene in the presence of freshly emitted diesel soot exhaust. *Atmos. Environ.* 2004, 38, 2597-2605.

[44] Sage, A.M.; Weitkamp, E.A.; Robinson, A.; Donahue, N.M. Evolving mass spectra of the oxidized component of organic aerosol: Results from aerosol mass spectrometer analyses of aged diesel emissions. *Atmos. Chem. Phys.* 2008, 8, 1139–1152.

[45] Weitkamp, E.A.; Sage, A.M.; Pierce, J.R.; Donahue, N.M.; Robinson, A.L. Organic aerosol formation from photochemical oxidation of diesel exhaust in a smog chamber. *Environ. Sci. Technol.* 2007, 41, 6969-6975.

[46] Weitkamp, E.A.; Lambe, A.T.; Donahue, N.M.; Robinson, A.L. Laboratory measurements of the heterogeneous oxidation of condensed-phase organic molecular makers for motor vehicle exhaust. *Environ. Sci. Technol.* 2008, 42, 7950–7956.

[47] Lambe, A.T. ; Miracolo, M.A.; Hennigan, C.J. ; Robinson, A.L.; Donahue, N.M. Effective Rate Constants and Uptake Coefficients for the Reactions of Organic Molecular Markers (n-Alkanes, Hopanes, and Steranes) in Motor Oil and Diesel Primary Organic Aerosols with Hydroxyl Radicals. *Environ. Sci. Technol.* 2009, 43, 8794-8800.

[48] Saathoff, H.; Moehler, O.; Schurath, U.; Kamm, S.; Dippel, B.; Mihelcic, D. The AIDA soot aerosol characterisation campaign 1999. *J. Aerosol Sci.* 2003, 34, 1277-1296.

[49] Geiger, H.; Kleffmann, J.; Wiesen, P. Smog chamber studies on the influence of diesel exhaust on photosmog formation. Atmos. Environ. 2002, 36, 1737–1747.

[50] Borrás, E.; Tortajada-Genaro, L.A.; Vázquez, M.; Zielinska, B. Polycyclic aromatic hydrocarbon exhaust emissions from different reformulated diesel fuels and engine operating conditions. *Atmos. Environ.* 2009, 43, 5944 -5952.

[51] Desantes, J.M.; Bermúdez, V.; García, J.M.; Fuentes, E. Effects of current engine strategies on the exhaust aerosol particle size distribution from a Heavy-Duty Diesel Engine. *J. Aerosol Sci.* 2005, 36, 1251-1276.

[52] Ravindra, K.; Sokhi, R.; Van Grieken, R. Atmospheric polycyclic aromatic hydrocarbons: Source attribution, emission factors and regulation. *Atmos. Environ.* 2008, 42, 2898-2921.

[53] Kubátová, A.; Steckler, T.S.; Gallagher, J.R.; Hawthorne, S.B.; Picklo, M.J. Toxicity of wide-range polarity fractions from wood smoke and diesel exhaust particulate obtained using hot pressurized water. *Environ. Toxicol. Chem.* 2004, 23, 2243-2250.

[54] Christensen, A.; Östman, C.; Westerholm, R. Ultrasound-assisted extraction and on-line LC–GC–MS for determination of polycyclic aromatic hydrocarbons (PAH) in urban dust and diesel particulate matter. *Anal. Bioanal. Chem.* 2005, 381, 1206-1216.

[55] Huang, W.; Smith Thomas, J.; Long, N.; Wang, T.; Chen, H.; Wu, F.; Herrick, R.F.; Christiani, D.C.; Ding, H. Characterizing and Biological Monitoring of Polycyclic Aromatic Hydrocarbons in Exposures to Diesel Exhaust. *Environ. Sci. Technol.* 2007, 41, 2711-2716.

[56] Riddle, S.G.; Jakober, C.A.; Robert, M.A.; Cahill, T.M.; Charles, M.J.; Kleeman, M.J. Large PAHs detected in fine particulate matter emitted from light-duty gasoline vehicles. *Atmos. Environ.* 2007, 41, 8658-8668.

In: Fourier Transform Infrared Spectroscopy
Editor: Oliver J. Rees, pp. 139-158

ISBN: 978-1-61668-835-6
© 2010 Nova Science Publishers, Inc.

Chapter 6

FOURIER TRANSFORM INFRARED SPECTROSCOPY: DEVELOPMENTS, TECHNIQUES AND APPLICATIONS

Noureddine Abidi *

Fiber and Biopolymer Research Institute, Department of Plant and
Soil Science, Texas Tech University, USA

ABSTRACT

The Universal Attenuated Total Reflectance Fourier Transform Infrared (UATR-FTIR) equipped with a ZnSe-Diamond composite crystal allows collection of FTIR spectra directly on a sample without any special preparation. The instrument is equipped with a "pressure arm" which is used to apply a constant pressure to the cotton samples positioned on top of the ZnSe-Diamond crystal to ensure a good contact between the sample and the incident IR beam and prevent the loss of the IR beam. The amount of pressure applied is monitored by the Perkin-Elmer FTIR software. In this chapter a review of applications of the FTIR to study cotton fibers. The UATR-FTIR is used to investigate the structural changes that occur during the different developmental stages of cotton fibers from fibers initiation to maturity. The UATR-FTIR is used also to analyze contaminated cotton fibers. In this case, the Principal Component Analysis of the FTIR spectra showed that it is possible to discriminate between contaminated and non-contaminated cotton. The results obtained showed that the ZnSe-Diamond composite crystal FTIR accessory could be used as a fast and precise nondestructive method to discriminate between contaminated and non-contaminated cotton. Furthermore, the UATR-FTIR is routinely used to investigate the cellulose chemical functionalization to impart different properties to cotton fabric.

1. INTRODUCTION

Fourier Transform Infrared spectroscopy (FTIR) has emerged as a key technique for the study of plant growth and development (McCann et al 2007; Yong et al 2005; Carpita et al

---

* Email: n.abidi@ttu.edu. Tel: 806- 742-5333

2001; Zeier and Schreiber 1999; Chen et al 1998; McCann et al 1997; Séné et al 1994; McCann et al 1993; McCann et al 1992). Zeier and Schreiber (1999) used FTIR to characterize isolated endodermal cell walls from plant roots and assigned FTIR frequencies to functional groups present in the cell wall, including the relative amounts of the cell wall biopolymers suberin and lignin, as well as cell wall carbohydrates and proteins. FTIR absorption spectra indicated structural differences for three developmental stages of the endodermal cell wall under study. The authors concluded that FTIR could be used as a direct and non-destructive method suitable for the rapid investigation of isolated plant cell walls. The approach has since been successfully applied to screen large numbers of mutants for a broad range of cell wall phenotypes using FTIR of leaves of *Arabidopsis thaliana* and flax (*Linum usitatissimum*) (Chen et al. 1998). In this study, Chen and co-workers reported that principal component analysis (PCA) of FTIR spectra can distinguish between mutants that are deficient in cell wall sugars. Also, FTIR and FT-Raman spectroscopy have been successfully used to investigate the primary cell wall architecture at a molecular level (Séné et al. 1994). Dynamic changes in cell wall composition of hybrid maize coleoptiles (*Zea mays*) were investigated by FTIR (McCann et al 2007). The authors reported that neural network algorithms could correctly classify infrared spectra from cell walls harvested from individuals differing at one-half-day interval of growth.

Furthermore, the Infrared spectroscopy has been used to confirm the effectiveness of the reactions between the crosslinking agent and the OH groups of the cellulose polymer. Morris, et al. (1994) used Near Infrared for quantitative determination of polycarboxylic acids on cotton fabrics. The method used requires grinding cotton fabric in a Wiley mill to pass a 20-mesh screen and pressing about 1.8mg of cotton fabric with 350 mg of KBr (potassium bromide). The carbonyl absorbance around 1700-1750 cm^{-1} was used to quantify the amount of polycarboxylic acids on cotton fabric. Wei, et al. (1999) used infrared spectroscopy as a tool for predicting the performance of durable-press-finished cotton fabric. The method used for preparing the sample for infrared measurements required grinding the cotton fabric into powder using a Wiley mill to improve the sample uniformity. However, these methods are destructive, labor intensive, and require a skilled operator in order to get satisfactory results. Abidi et al. (2005) used the Universal Attenuated Total Reflectance Fourier Transform Infrared (UATR-FTIR) with a ZnSe-Diamond composite crystal to analyze chemical cross-linked cotton fabric.

2. UNIVERSAL ATTENUATED TOTAL REFLECTANCE FOURIER TRANSFORM SPECTROSCOPY

The FTIR spectra of cotton fiber samples were recorded in an environmentally-controlled laboratory maintained at relative humidity of 65±2% and 21±1°C using the Spectrum-One equipped with an UATR (Universal Attenuated Total Reflectance) accessory (Perkin-Elmer, USA). The UATR-FTIR was equipped with a ZnSe-Diamond crystal composite that allows collection of FTIR spectra directly on a sample without any special preparation. The instrument is equipped with a "pressure arm" which is used to apply a constant pressure to the cotton samples positioned on top of the ZnSe-Diamond crystal to ensure a good contact between the sample and the incident IR beam and prevent the loss of the IR beam. The amount of pressure applied is monitored by the Perkin-Elmer FTIR software.

Thirty FTIR spectra per sample were acquired for each developmental stage to produce a total of 180 spectra (30 spectra x 3 replications per dpa x 2 greenhouse replications). All FTIR spectra were collected at a spectrum resolution of 4 cm^{-1}, with 32 co-added scans over the range from 4000 cm^{-1} to 650 cm^{-1}. A background scan of clean ZnSe-Diamond crystal was acquired before scanning the samples.

The Perkin-Elmer software was used to perform spectra normalization, baseline corrections, and peak integration. FTIR spectra were then exported to Excel and were subjected to Principal Component Analysis (PCA) with leverage correction and mean-center cross validation boxes checked using Unscrambler V. 9.6 Camo Software AS (CAMO Software AS, Norway).

2.1. FTIR Study of Cotton Fiber Development

Cotton fiber development consists of five major overlapping developmental stages (Wilkins and Jernstedt 1999): differentiation, initiation, polar elongation, secondary cell wall deposition, and maturation. The day of flowering is referred to as anthesis and the term "days post-anthesis" (dpa) is often used to describe the cotton fiber development. Fiber initiation, which commences at 0 dpa, signals the onset of fiber morphogenesis. Fiber growth is characterized by the synthesis of the primary cell wall and an increase in fiber length up to ~30 mm within 3 weeks after anthesis. The stage of secondary cell wall development commences in general around 21 dpa and continues for a period of ~3 to 6 weeks post-anthesis. This phase is marked by a massive deposition of a thick cellulosic wall (Wilkin and Jernstedt 1999). The transition period between 16 and 21 dpa is considered to represent a developmental switch in emphasis from primary to secondary cell synthesis during cotton fiber development. During these developmental stages, important structural changes occur leading to cellulose macromolecules formation ($\beta(1\rightarrow4)$ glucopyranose).

The UATR-FTIR was used to investigate the structure and composition of fiber during different phases of development of cotton fibers. For this purpose, two independent replications (10 plants each) of two cotton cultivars (*Gossypium hirsutum* L. cv. TX19 and TX55) were planted in a greenhouse with day/night cycles varying from 13/11 to 11/13 hours and day/night temperatures of about 31°C / 24°C. Plants were grown in 20 liters (5 gallons) pots of Sungrow SB 300 potting mix that had been amended with Peters 15-9-12 slow release fertilizer prior to potting. Plants were watered as needed. On the day of flowering (0 dpa), individual flowers were tagged, and 14 developing bolls per cultivar and per replication were harvested at 10, 14, 17, 18, 19, 20, 21, 24, 27, 30, 36, 46, and 56 dpa. The pericarp was immediately removed (excised with scalpel) and isolated ovules were transferred into cryogenic vials and stored in a Cryobiological Storage System filled with liquid nitrogen for analyses.

Frozen cotton fiber samples were dehydrated using the procedure described in (Abidi et al. 2008; Muller and Jackset al. 1975; Rajasekaran et al. 2006). Frozen samples were first rinsed with water and then washed with acidified solution of 2,2-dimethoxypropane (one drop of HCl in 50 ml of 2,2-dimethoxypropane), followed by five exchanges for 15 minutes each in 100% acetone. In a slightly acidic solution, 2,2-dimethoxypropane is instantly hydrolyzed by water to form methanol and acetone (Muller and Jacks 1975).

Figures 1-a and 1-b shows a series of FTIR spectra of fibers from TX19 and TX55 cultivars respectively at 14, 18, and 21 dpa. Compared to the FTIR spectra of mature cotton fibers (Figure 1-c), several additional vibration bands are noticed in the FTIR spectra of developing cotton fibers.

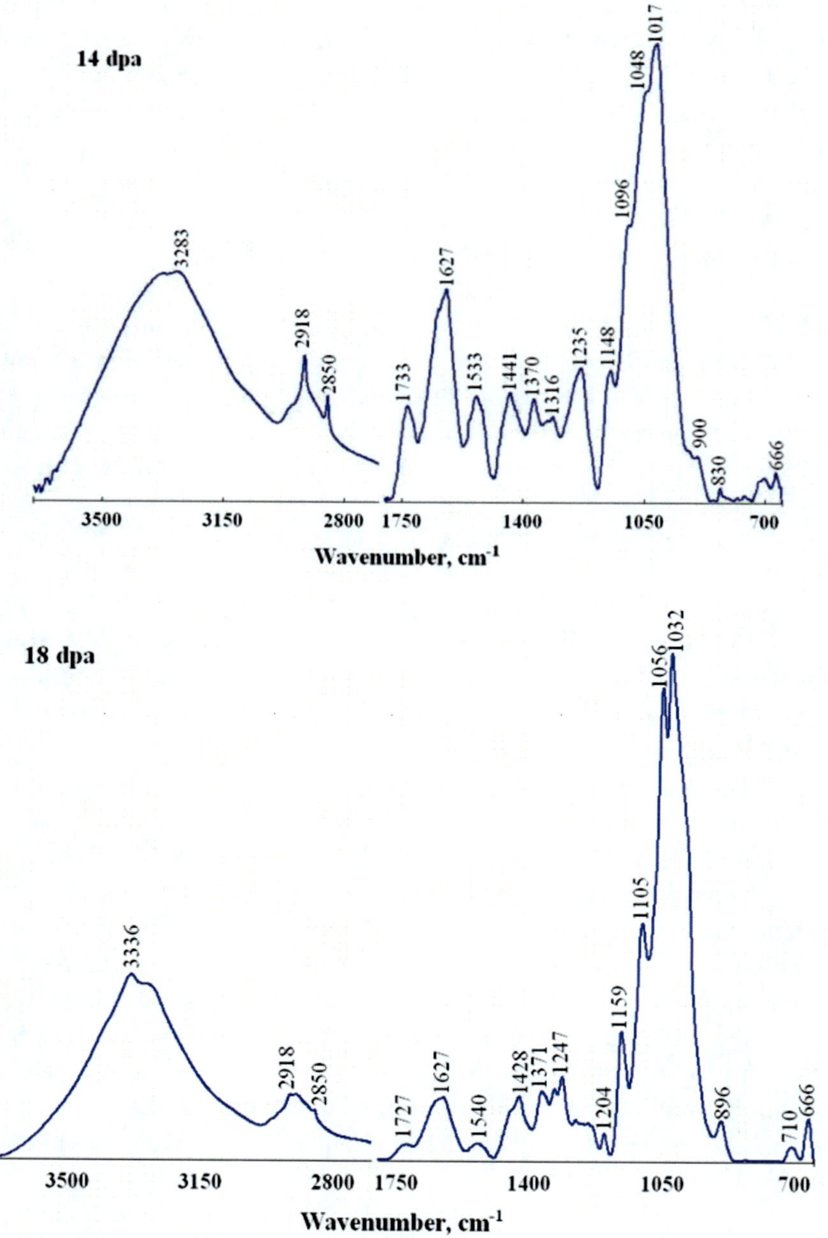

Figure 1-a. (Continued).

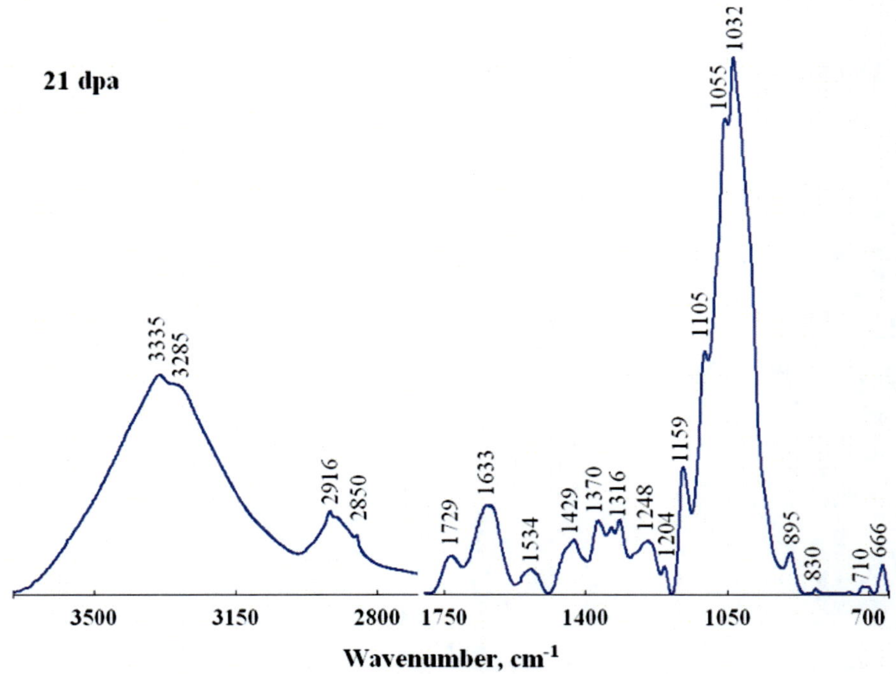

Figure 1-a. FTIR spectra of developing cotton fibers (*Gossypium hirsutum* L. cv. TX19) at different days post-anthesis (dpa).

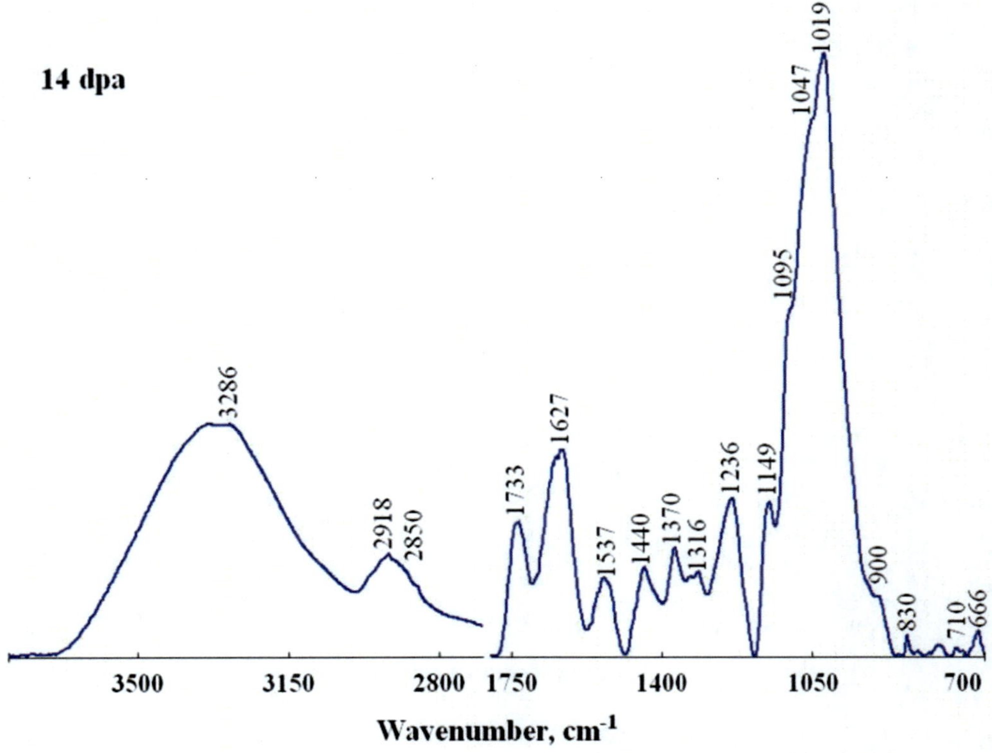

Figure 1-b (Continued)

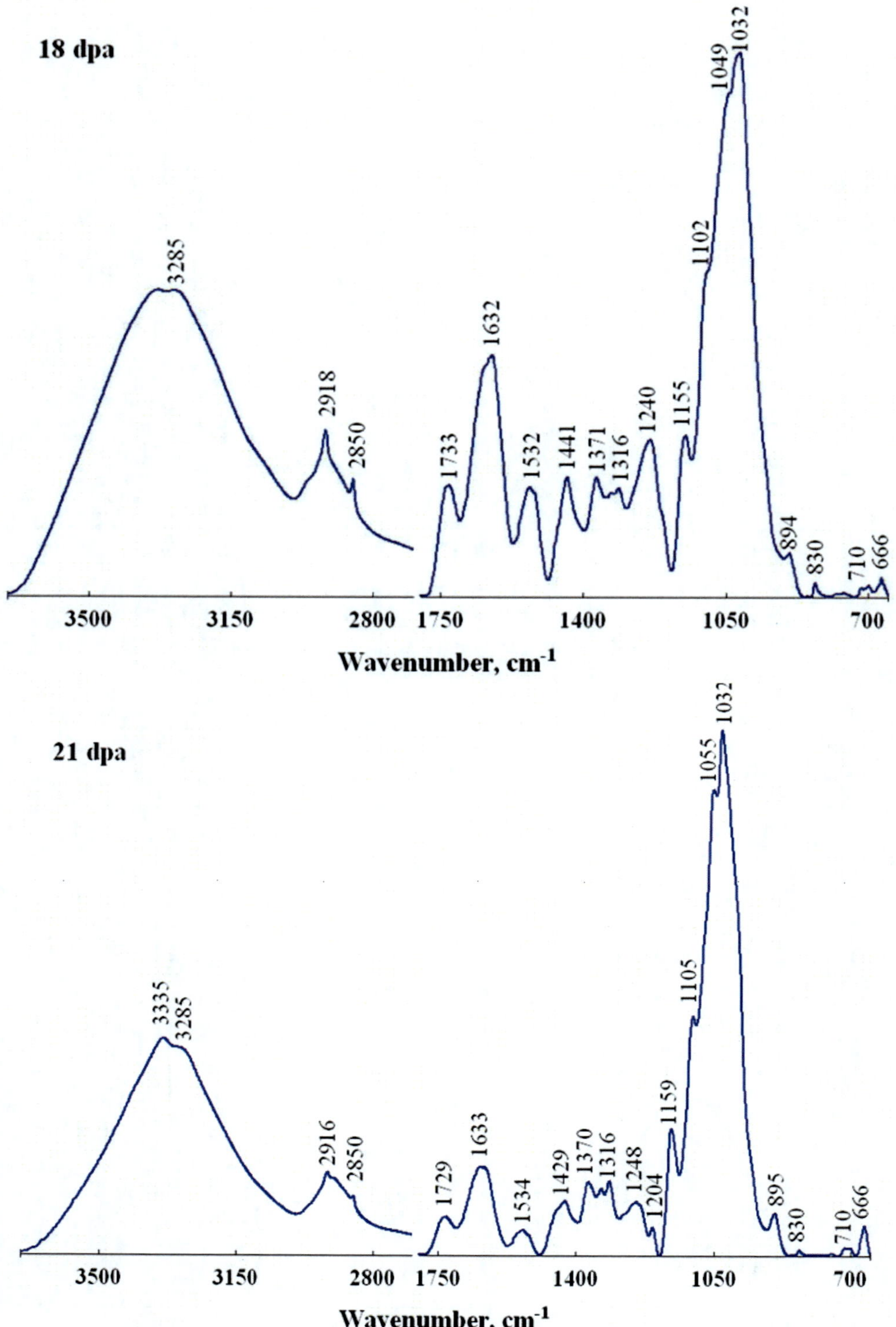

Figure 1-b. FTIR spectra of developing cotton fibers (*Gossypium hirsutum* L. cv. TX559) at different days post-anthesis (dpa).

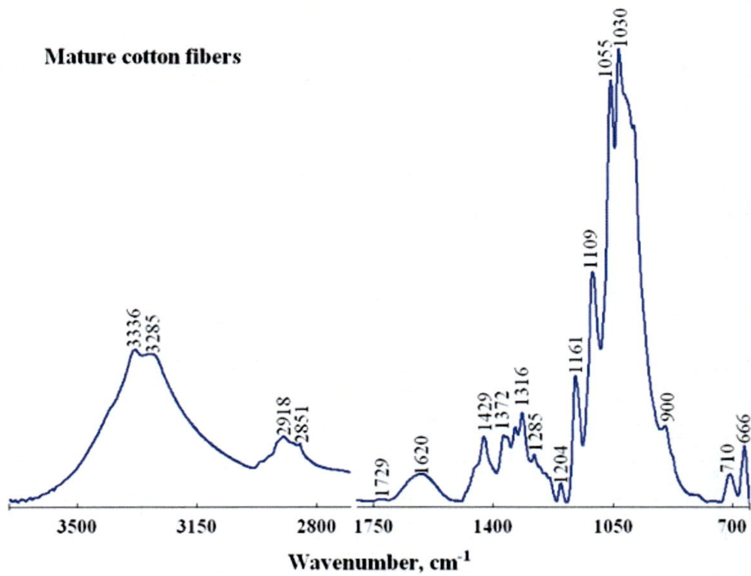

Figure 1-c. FTIR spectrum of mature cotton fibers.

Vibrations located at 2918 and 2850 cm⁻¹: these vibrations are attributed to –CH₂ asymmetric vibrations and could originate from the presence of wax substances present on the surface of the primary cell wall (Abidi et al. 2008 ; Abidi et al. 2010). The intensities of these two peaks start decreasing at 19 dpa. This result is in agreement with our previous findings (Abidi et al. 2008 ; Abidi et al. 2010). We reported that the percent contribution of the primary cell wall to the total weight of the fiber decreased as wall thickness increases (increased fiber maturity, thus secondary cell wall). Therefore, the relative importance of the vibration bands attributed to noncellulosic substances (e.g. waxes which are located essentially on the primary cell wall) is less.

Vibration located at 1733 cm⁻¹: this vibration is attributed to C=O stretching vibration and could originate from esters or amides (Abidi et al. 2008 ; Abidi et al. 2010). The integrated intensity of this peak (I_{1733}) was calculated between 1780 cm⁻¹ and 1701 cm⁻¹ and was reported as function of dpa. The results obtained showed that I_{1733} decreased continuously between 10 and 24 dpa. These results indicate structural changes are occurring during fiber development.

Vibration located at 1627 cm⁻¹: this vibration is attributed to O-H bending of adsorbed water molecules (Abidi et al. 2008 ; Abidi et al. 2010). The integrated intensity of this peak (I_{1627}) was calculated between 1701 cm⁻¹ and 1576 cm⁻¹. The change in the integrated intensity I_{1627} as function of dpa showed that the amount of adsorbed water decreased linearly until the fibers reached 24 dpa. The decrease in the amount of adsorbed water could be attributed to the decrease of the surface area and to reduced accessibility of water molecules to the internal hydroxyl groups to establish hydrogen bonding. This could be the result of increased polymerization reactions of glucose units to form cellulose macromolecules followed by increased crystallinity. Hsieh et al. reported that the degree of crystallinity of two cotton fiber cultivars (Maxxa and SJ-2) increases beginning 24 dpa (Hsieh et al. 1997). The results indicated that the major change in crystallinity occurred between 24 dpa and 28 dpa, and no significant changes occurred thereafter. These results are in agreement with our FTIR results, which indicate that no change in the amount of adsorbed water is noticed between 27 dpa and 56 dpa. In this developmental stage, fibers from both cultivars have nearly the same amount of adsorbed water. This could indicate that the remaining water molecules are those which are strongly bonded to cellulose macromolecules via hydrogen bonding.

Vibration located at 1534 cm^{-1}: this vibration is attributed to NH_2 deformation and likely indicative of proteins or amino acids (Abidi et al. 2008 ; Abidi et al. 2010). The integrated intensity of this peak (I_{1534}) was calculated between 1575 cm^{-1} and 1487 cm^{-1} and the evolution of I_{1534} as function of dpa showed a behavior similar to I_{1733}. A continuous decrease between 10 and 24 dpa is observed.

Vibrations located at 1236 and 1204 cm^{-1}: the vibration located at 1236 cm^{-1} is attributed to C=O stretching or NH_2 deformation (Abidi et al. 2008 ; Abidi et al. 2010). This vibration decreased in intensity during fiber development. The disappearance of this band is accompanied by the appearance of a vibration at 1204 cm^{-1} at 20 dpa. This vibration has been attributed by Ilharco et al. to C-O-C stretching mode of the pyranose ring (Ilharco et al. 1997). This vibration appears almost simultaneously with the vibration located at 900 cm^{-1} (attributed to β-linkage). Consequently, these two vibrations could be considered as the secondary cell wall's fingerprint.

Vibrations located at 1148, 1096, and 1048 cm^{-1}: these vibrations are present only as shoulders in the spectra of fibers of age between 10 and 19 dpa and become sharper beginning at 20 dpa. It is also noteworthy the shift at 20 dpa of the vibration 1148 cm^{-1} to 1159 cm^{-1}, the vibration 1096 cm^{-1} to 1105 cm^{-1}, and the vibration 1048 cm^{-1} to 1056 cm^{-1}. The vibration located at 1161 cm^{-1} is assigned to the anti-symmetric bridge C-O-C stretching vibration (Ilharco et al. 1997). The vibration located at 1105 cm^{-1} is assigned to anti-symmetric in-plane ring stretching band (Ilharco et al. 1997). The vibration located at 1056 cm^{-1} is attributed to C-O stretching mode (Abidi et al 2008).

Vibrations located at 1017, 1031, 103, 985, and 900 cm^{-1}: the vibrations located at 1017 cm^{-1}, attributed to C-O stretch, is shifted to 1031 cm^{-1} at 19 dpa. The vibrations located at 1003 cm^{-1} and 985 cm^{-1} appeared at 56 dpa. These vibrations are attributed to C-O and ring stretching modes (Liang and Marchessault 1959). The vibration located at 900 cm^{-1} is attributed to β-linkage (Abidi et al. 2008). This vibration is present only as a small shoulder in the FTIR spectra of fibers at 10 dpa but becomes sharper in the FTIR spectra of fibers at 19 dpa.

2.2. FTIR Study of Cotton Fiber Contamination

Cotton stickiness caused by excess sugars on the lint, from the plant itself (Hendrix et al, 1995) or from insects (Sisma and Scheneck, 1984) is a very serious problem that affects all segments of the cotton industry (Hequet et al, 2000 ; Watson, 2000). Stickiness is a worldwide contamination problem with around one fifth of the world production affected to some degree (Strolz, 2002). The sugar composition of water extracts from honeydew-contaminated cottons analyzed by High Performance Liquid Chromatography (HPLC) is extremely complex. Among the identifiable sugars, the disaccharide trehalulose and the trisaccharide melezitose are specific to insects and are not found in the plant. In general, a high percentage of melezitose along with a low percentage of trehalulose reveals the presence of aphid (*Aphis gossypii* Glover) honeydew. Conversely, a dominance of trehalulose indicates contamination by whitefly (*Bemisia argentifolii* Bellows and Perring [= *B. tabaci* (Gennadius) strain B]) honeydew. Hendrix et al. (Hendrix et al, 1992) analyzed the honeydew from insects and found that the aphid honeydew contained around 38.3% melezitose plus 1.1% trehalulose, while the whitefly honeydew contained 43.8% trehalulose plus 16.8% melezitose. However, these relative percentages will vary depending on the environmental or feeding conditions. In other studies, it has been demonstrated that the severity of stickiness incidents in the textile mill is related to the type of sugars present on the lint (Miller et al, 1994).

The UATR-FTIR is very useful technique to analyze cotton contaminations and to discriminate between sticky and non sticky cotton. For this application, seventeen commercial cotton bales contaminated with insect honeydew to different degrees were selected. One commercial non sticky cotton bale was selected as a control. Ten samples (100 g each) were taken from each bale. All samples were conditioned for at least 96 h in standard laboratory conditions at $65 \pm 2\%$ Relative Humidity and $21 \pm 1°C$ before testing. The sugars present on the contaminated lint were identified and quantified using High Performance Liquid Chromatography (Dionex Corporation, Sunnyvalle, CA) (Abidi et al 2007). It has been reported that trehalulose is the primary sugar causing cotton contamination. Figure 2 shows the FTIR spectra of non contaminated cotton fibers and hydrated trehalulose. As shown in this figure, the peaks located at 3280 cm^{-1}, 1622 cm^{-1}, and 1018 cm^{-1} are very intense in the spectrum of hydrated trehalulose. These peaks are attributed, respectively to, O-H stretching vibration of absorbed water molecules, O-H bending vibration in H_2O, and C-O stretching vibration in the primary alcohol ($-H_2C-OH$). In addition, the peaks located at 919 cm^{-1}, 867 cm^{-1}, 823 cm^{-1}, and 779 cm^{-1} are present in the spectra of hydrated trehalulose but not in the spectra of cotton fibers.

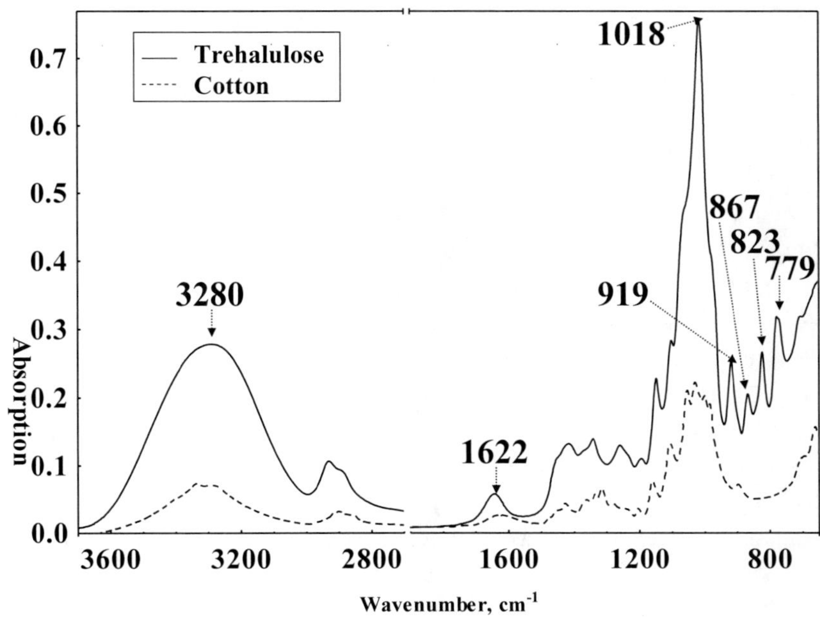

Figure 2. FTIR spectra: trehalulose versus cotton fibers.

The hygroscopic properties of trehalulose were studied by observing the changes in the FTIR peak intensities of the vibration bands located at 3280 cm^{-1} and 1018 cm^{-1} (Abidi et al 2005). It was found that the integrated intensities of these peaks increased with increasing hydration time of trehalulose. Because of the high hygroscopicity of trehalulose, these two peaks could be used in combination with the peak located at 1622 cm^{-1} to discriminate between whitefly honeydew contaminated cotton and non-contaminated cotton.

The integrated intensities of these peaks were calculated (respectively I_{3280}, I_{1622}, and I_{1018}). In order to obtain the integrated intensities, the peak located at 3280 cm^{-1} was integrated from 3000 cm^{-1} to 3700 cm^{-1}, the peak located at 1622 cm^{-1} was integrated from 1725 cm^{-1} to 1530 cm^{-1}, and the peak located at 1018 cm^{-1} was integrated from 1095 cm^{-1} to 941 cm^{-1}. These integrated intensities were correlated with the percentage of sugars identified on the contaminated cotton fibers. Only trehalulose content showed a very significant correlation.

Integrated intensities of the peaks located at 3280, 1622, and 1018 cm^{-1} were calculated as function of the percentage of trehalulose in the total identified. On average, for cotton contaminated with whitefly honeydew (%Trehalulose is comprised between 25% and 45% on the total identified sugars), the integrated intensity of the peak at 3280 cm^{-1} is 7 times higher than for non-contaminated cotton. The following relationships were established between the integrated intensities and the trehalulose content:

- Peak 3280 cm^{-1}: $I_{3280} = 7.172 + 2.764*(\%trehalulose) - 0.029*(\%trehalulose)^2$, Adjusted $R^2 = 0.93$, $F(2,15) = 113.99$, $p < 0.001$, Standard error of estimate = 4.7169.
- Peak 1622 cm^{-1}: $I_{1622} = 0.408 + 0.410*(\%trehalulose) - 0.005*(\%trehalulose)^2$, Adjusted $R^2 = 0.92$, $F(2,15) = 93.48$, $p < 0.001$, Standard error of estimate = 0.6510.
- Peak 1018 cm^{-1}: $I_{1018} = 7.106 + 0.512*(\%trehalulose)$, Adjusted $R^2 = 0.82$, $F(1,16) = 79.30$, $p < 0.001$, Standard error of estimate = 2.6308.

These results indicate that quantification of trehalulose is possible by means of FTIR at selected wavenumbers. Nevertheless, the FTIR spectra could contain other pertinent information that could allow discriminating between contaminated and non-contaminated cotton. Indeed, the composition of insect honeydew is quite complex. On average, 77% of all sugars present in honeydew are identified with HPLC (Hequet and Abidi, 2006). The largest polysaccharides present in honeydew are not identified by HPLC while they are known to contribute to the stickiness phenomenon (Hequet and Abidi, 2006 ; Wei et al 1997). Therefore, an exploratory Principal Component Analysis (PCA) was carried out in order to identify distinct groups of FTIR spectra. PCA technique is a good method to use for data exploration. It rearranges information in such a way that patterns may be revealed, which, although present, were obscured in the original data. Four representative samples were selected: Sample 3103 represented non contaminated cotton (control), sample 3139 was heavily contaminated cotton with whitefly honeydew, sample 2124 was contaminated with whitefly and aphid honeydew, sample 2391: was slightly contaminated with aphid honeydew.

Binary comparisons between contaminated cotton and non contaminated cotton were done. The plots of the PC1 versus PC2 scores are depicted in figures 3-a to 3-c. The first principal component axis (PC1) allows to discriminate between the FTIR spectra of contaminated cotton and non-contaminated cotton, the later having negative scores. The FTIR spectra in figures 3-b to 3-c fall into two very distinct clusters corresponding to contaminated and non-contaminated cotton. In figure 3-a, however, there is some overlap between the FTIR spectra of contaminated cotton and non-contaminated cotton. This is because cotton 2391, is only slightly contaminated with aphid honeydew. Thus, the probability of finding a sticky point with the UATR-FTIR is low. This translates into an overlap of FTIR spectra from the non-contaminated parts of the sample.

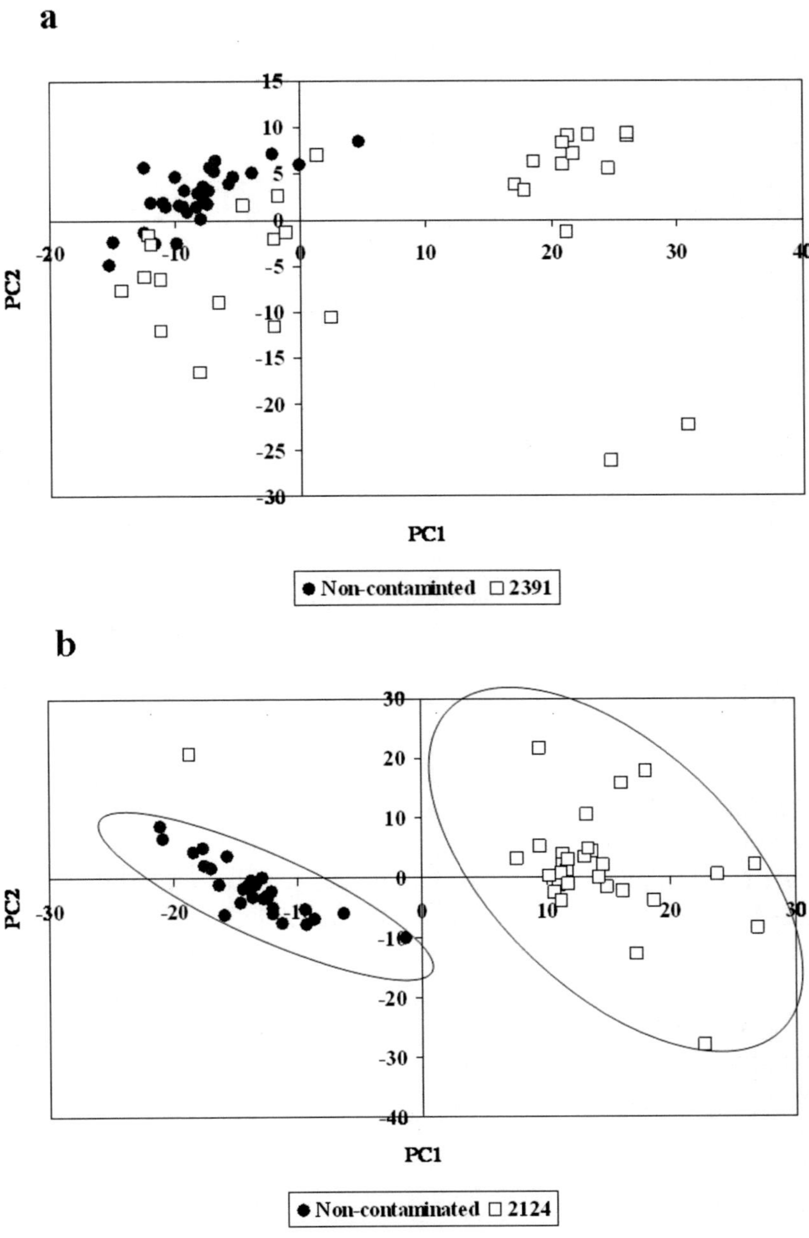

Figure 3. (Continued).

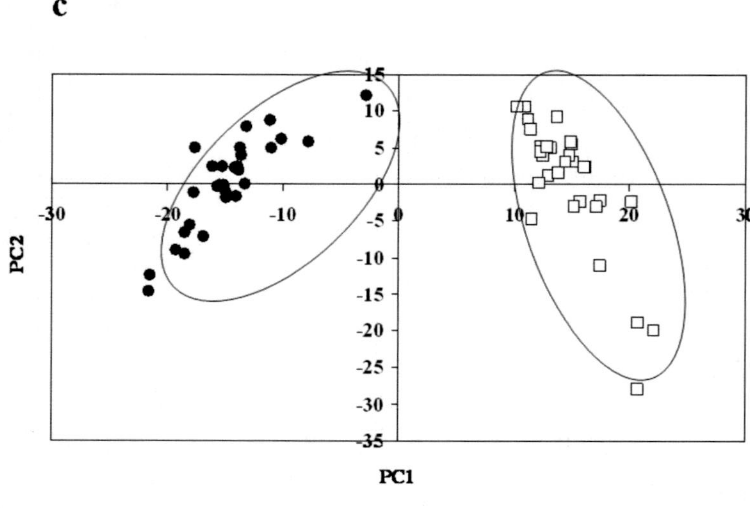

• **Non-contaminated** □ **3139**

Figure 3. Principal Component Analysis of FTIR spectra in the 4000-650 cm^{-1} range. (a) Non-contaminated versus cotton 2391, (b) Non-contaminated versus 2124, (c) Non-contaminated versus 3139.

2.3. UATR-FTIR Study of Chemical Cross-Linking of Cotton Fabric

Cotton is made up of cellulose macromolecules with repeating anhydroglucose units. On each unit, there are three available hydroxyl groups. These hydroxyl groups serve as sites for water molecules absorption by establishing many hydrogen bonds with the cellulose macromolecules. A severe limitation of fabrics made from cellulosic fibers is their tendency to wrinkle. In general, wrinkles occur when the fiber is bent. In this process, hydrogen bonds between the cellulose macromolecules in the amorphous regions of the fibers break, thus, allowing the chains to slip past one another. The hydrogen bonds then reform in new places and hold creases in the fiber and fabric.

The basic idea behind the resistance of cotton fabric to wrinkles is to restrict the slippage of cellulose chains (Perkins, 1996). Appropriate chemical treatment of cotton fabric enables to establish covalent links between the cellulosic chains in the amorphous regions of the fibers, thus, restricting the slippage of the cellulosic chains. For many years, the textile industry has been using N-methylol based products with very low formaldehyde release as the crosslinking agent.

The most common method for finishing the cotton fabric is the pad-dry-cure process (Abidi et al. 2005 ;Wei and Yang, 1999 ; Yang et al, 2001 ;Yang and Wei, 2000 ; Yang et al, 1998). This method consists of impregnating the sample in an aqueous solution containing the crosslinking agent and the appropriate catalyst, padding the impregnated fabric to 90-100% wet pick-up, drying, and then curing.

The UATR-FTIR with a ZnSe-diamond crystal was used to assess the chemical finishing of cotton fabrics. For this purpose, cotton fabrics were treated with increasing amounts of a textile-finishing agent (DMUG = 1,3-Dimethyl-4,5-dihydroxy-2-imidazolidinone) to impart durable press properties according to the protocol described in (Abidi et al, 2005).

FTIR Integrated Intensity vs. %DMUG

Figure 4 shows representative FTIR spectra of untreated and treated cotton fabric with increasing amount of DMUG. The comparison between the spectra shows the presence of an additional peak around 1710 cm^{-1} for treated fabrics. This band is attributed to $-C=O$ stretching vibrations and is indicative of the presence of DMUG on the treated fabric.

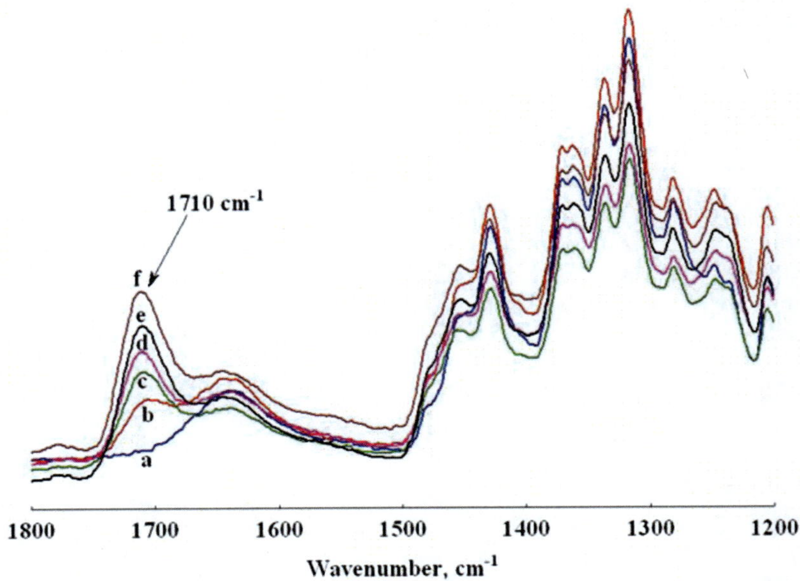

Figure 4. UATR-FTIR spectra of the control and the treated cotton fabrics. (a) Control, (b) 2%, (c) 10%, (d) 12%, (e) 15%, (f) 20%.

The vibration located around 1710 cm^{-1} was integrated from 1750 cm^{-1} to 1670 cm^{-1} to obtain the integrated intensity (I_{1710}) for each DMUG concentration and for each fabric. Figure 5 shows the plot of the integrated intensity versus the percent of DMUG initially in the crosslinking solution.

The non-linear relationships show high degrees of correlation between the concentration of the crosslinking agent DMUG in the solution and the concentration of the DMUG effectively establishing a cross-link between cellulose chains. The prediction equation is: I_{1710} = 458.2(%DMUG)^2 + 0.73, with adjusted R^2=0.96. The decreasing slope of the curve is due to the unavailability of cellulosic OH groups for crosslinking with the OH groups of the DMUG (saturation phenomenon). The FTIR measurements of the quantity of crosslinking agent were performed after the required 5 laundering and tumble-drying cycles.

The standard test for smoothness appearance was performed according to AATCC Test Method 124. This test method consists of 5 subsequent laundering and tumble-drying cycles. The treated fabrics were stitched to prevent unraveling and washed as directed in the AATCC TM 124. Laundering was conducted at a wash temperature of 41±1°C for 10 minutes, with 66±0.1g of AATCC standard detergent without optical brighteners. Tumble-drying was set for durable press conditions (30 minutes).

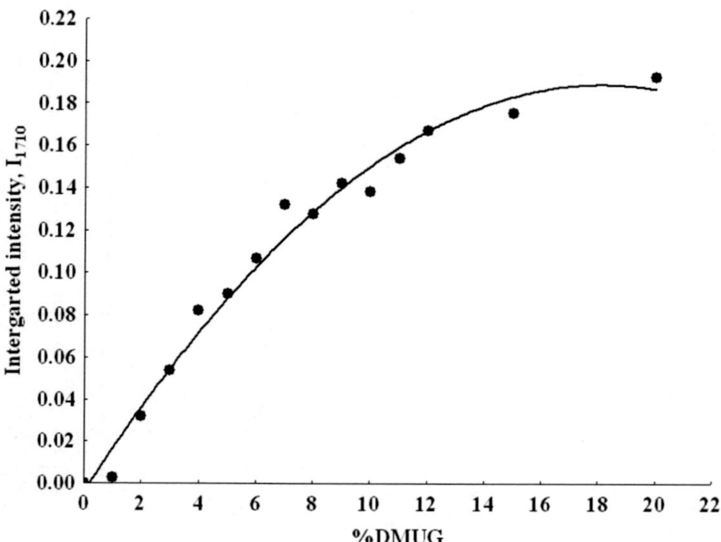

Figure 5. FTIR integrated intensity I_{1710} *vs.* %DMUG.

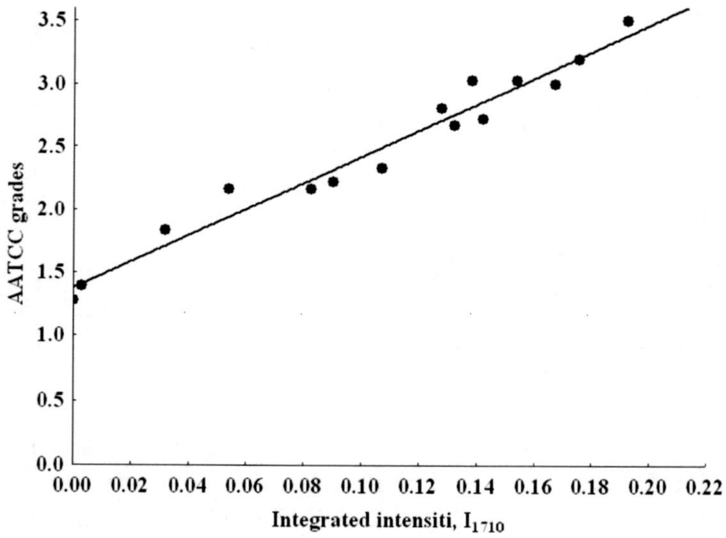

Figure 6. AATCC grade *vs.* FTIR integrated intensity I_{1710}.

At the end of the washing cycles, the individual samples were placed on perforated screen for conditioning at $65 \pm 2\%RH$ and $21 \pm 1°C$ for 24 hours, as directed by ASTM D1776 (Standard Practice for Conditioning and Testing Textiles). Three trained observers, using AATCC standard replicas, performed the smoothness appearance grading (also referred as Durable Press rating). All AATCC grading was performed before the FTIR measurements and the textile performance tests.

Figure 6 shows the relationships between the integrated intensity I_{1710} as measured with UATR-FTIR and the AATCC grades of the two fabrics. As expected, there is an increase of AATCC grades (i.e., smoother fabrics) with increasing DMUG concentrations.

3. FTIR IMAGING

Traditional Fourier Transform Infrared (FTIR) measurements have been made with devices capable of performing small sample area measurements such as individual sticky spot, seed coat fragment or individual fiber. However, with recent technological development FTIR devices can now image larger sample area.

For one sample, several spectra can be measured and combined to generate an IR image. FTIR mapping provides researchers with a unique tool for acquisition of spatially resolved molecular structure information. This technique has been applied to obtain 3D models of chemical structure information from serial 2D-IR images of human cortical bone (Ou-yang et al., 2002). Himmelsbach et al. (2003) used FTIR mapping and histochemical staining to reveal the location and relative importance of chemical components involved with the base of cotton fibers and their associated seed coat. The results established the usefulness of the FTIR.

We used the FTIR micro-spectroscopy for mapping and detection of cotton stickiness. FTIR spectra are automatically collected form specified location of the cotton fiber sample and IR images (false-color map) are generated.

IR images were acquired with the Autoimage FT-IR micro-spectroscopy system (from PerkinElmer, Norwalk, CT). The microscope includes a camera and a viewing system that magnifies the visible-light image of the sample and enables isolating a region of interest. The image of the sample is displayed on the monitor visible window.

The AutoIMAGE software (version 5.0.0 B8) enables the control of the operation of the microscope and maps and collects the spectra from a sample. The system used a KBr beam splitter and utilized a liquid nitrogen cooled MCT detector housed in the microscope. The Autoimage was further equipped with an automated XYZ motorized stage that was operated in the auto-focusing (Z-direction) mode under the software control. Both visible images and infrared maps were obtained in the transmission mode. Cotton fibers were transferred to the BaF_2 window (13 mm diameter, 2 mm thickness). A background spectrum of clean BaF_2 surface was collected before each scan. The IR spectra were recorded using 128 scans with 2 cm^{-1} spectral resolution between 4000 cm^{-1} and 700 cm^{-1}.

Cotton fibers were first visualized on the monitor and a visible image was recorded. The FTIR map scan enables to define a region of interest from the sample and to scan at regularly-spaced points. Chemical profiles were produced from the mapping procedure by extracting and displaying specific absorptions on the basis of their band intensities from the entire map area to produce a profile from that band. False-color 2-D maps were generated from each scan.

Figure 2 shows the FTIR spectra of non-contaminated cotton fiber and hydrated trehalulose. The comparison of these spectra reveals that: (1) the vibration bands corresponding to hydrogen bond (~3280 cm^{-1}: O-H stretching in absorbed water) and to absorbed water (1622 cm^{-1}: O-H bending vibration of H_2O) molecules are very intense in the spectra of hydrated trehalulose, (2) the vibration bands located at 1018 cm^{-1}, 919 cm^{-1}, 867 cm^{-1}, 823 cm^{-1} and 779 cm^{-1} are very intense in the spectra of hydrated trehalulose. In addition, bands located around 919 cm^{-1}, 867 cm^{-1}, 823 cm^{-1} and 779 cm^{-1} are specific to trehalulose. Therefore, in the FTIR maps of contaminated cotton fibers with whitefly honeydew (trehalulose-rich honeydew), these bands are expected to be easily identified.

Figure 7 shows a visible image of contaminated cotton fibers with whitefly honeydew and the corresponding FTIR map (total absorbance). The red color region, associated with high absorbance, is indicative of localized honeydew contamination. Figure 8 shows an example of the FTIR map generated from another contaminated cotton fibers with whitefly honeydew.

The false-color 2D FTIR maps are based on the FTIR spectra collected. A FTIR spectrum is associated with each pixel of the map. Figure 9 shows the FTIR spectrum extracted from the map (X mark on the FTIR map). The absorbance in this region is between 0.60 and 0.90.

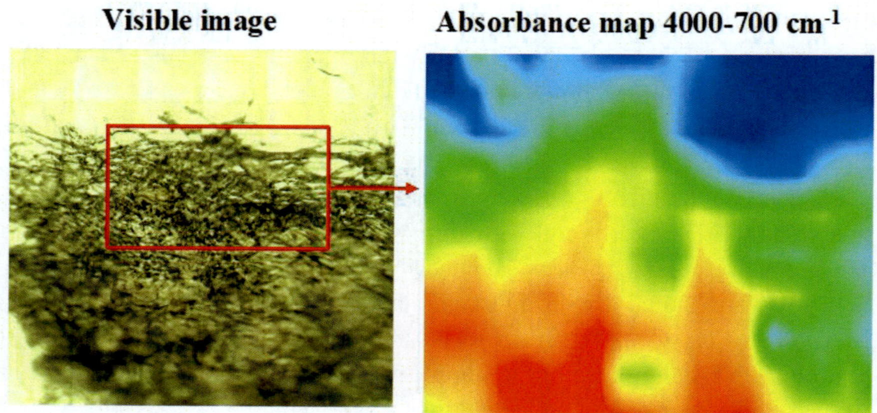

Figure 7. FTIR map of cotton fiber contaminated with whitefly honeydew – total absorbance 4000-700 cm⁻¹.

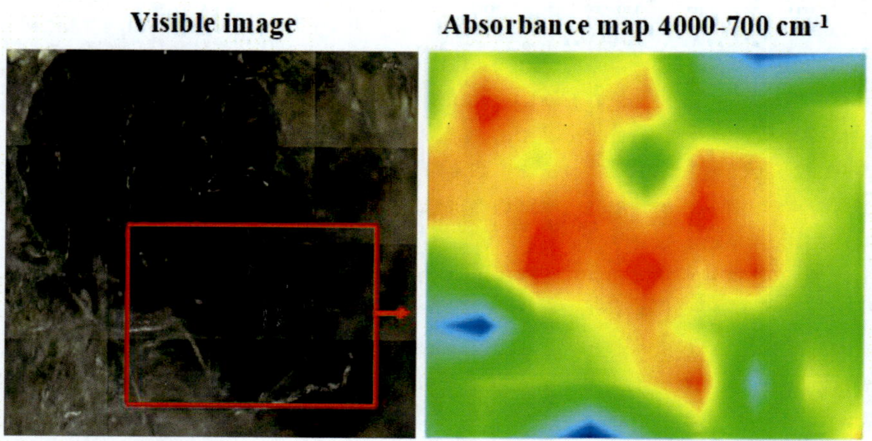

Figure 8. FTIR map of cotton fiber contaminated with whitefly honeydew – total absorbance 4000-700 cm⁻¹.

The comparison between the extracted spectrum and the spectrum of hydrated trehalulose shown in figure 2 indicates high concentration of water as well as the presence of trehalulose (presence of the vibrations in the fingerprint region of trehalulose) in the mapped contaminated cotton sample (Figure 10). However, in the spectra extracted from high absorbance (absorbance > 1.3) area in the map (Figure 11), the water vibration bands dominate the FTIR spectra.

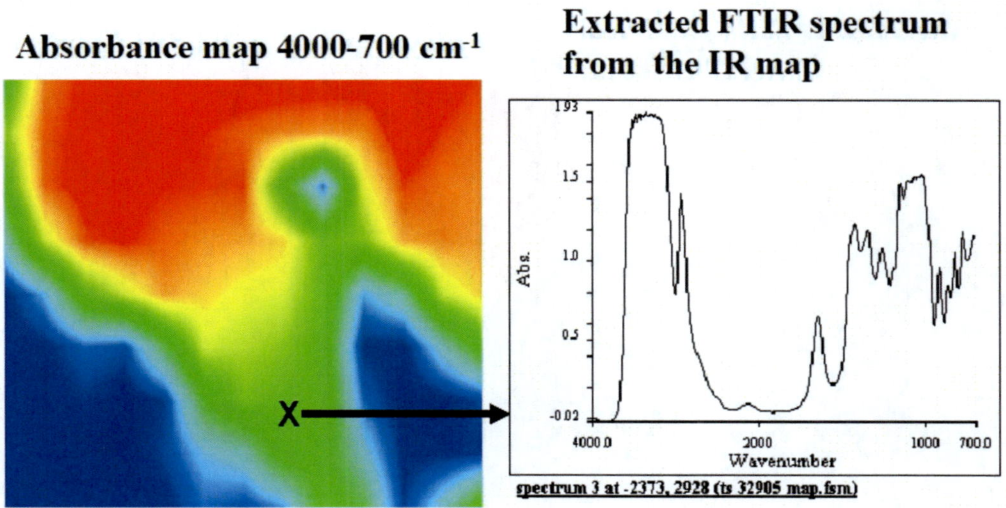

Figure 9. FTIR map of cotton fiber contaminated with whitefly honeydew and extracted FTIR spectrum from the position marked by "X".

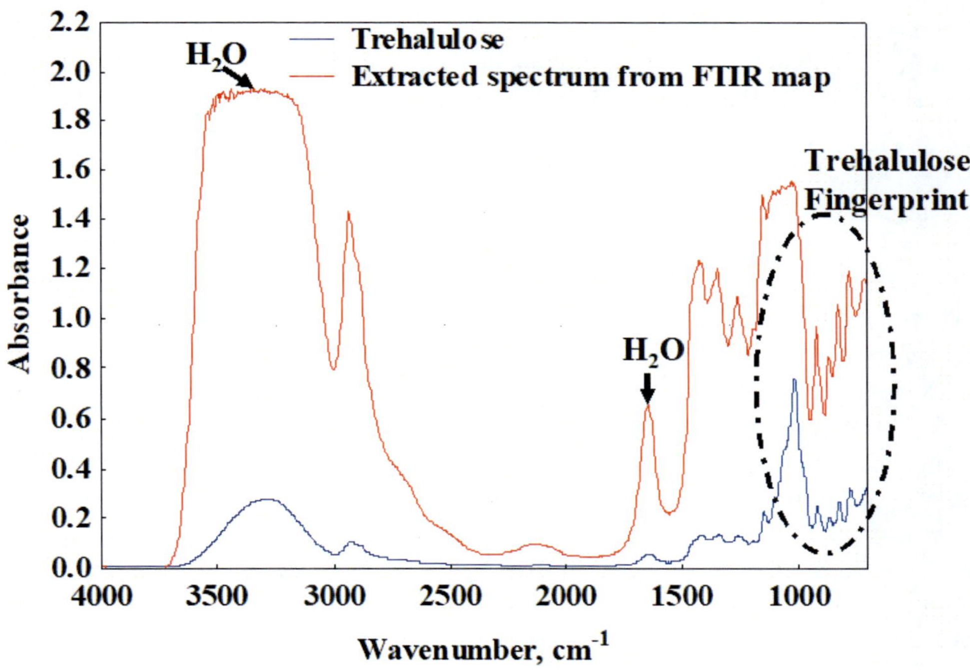

Figure 10. Comparison between the FTIR spectrum of hydrated trehalulose and extracted spectrum from the FTIR map.

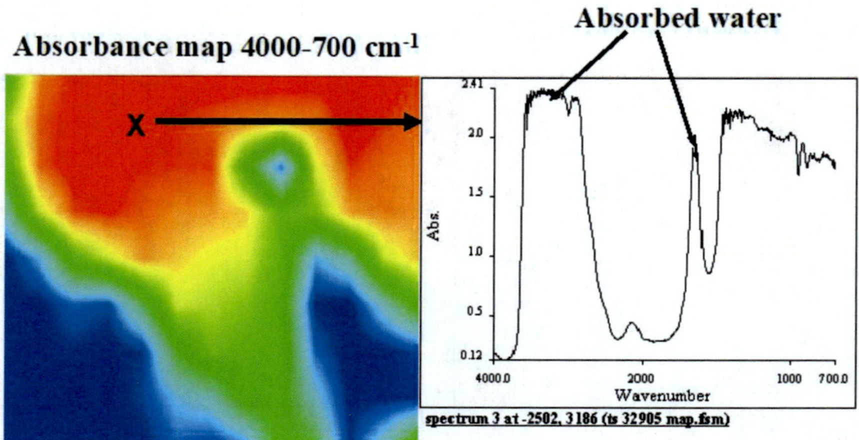

Figure 11. FTIR map of cotton fiber contaminated with whitefly honeydew and extracted FTIR spectrum from the position marked by "X".

In conclusion, FTIR mapping of contaminated cotton is a promising way to establish functional groups distribution. Regions with high concentrations of trehalulose show only the fingerprint of water, which masks the trehalulose fingerprint. Regions with low trehalulose concentration show less water, therefore, the trehalulose fingerprint is clearly visible. The high amount of adsorbed water is associated with the high hygroscopicity of trehalulose.

CONCLUSION

Fourier Transform Infrared microspectroscopy is proven to be very versatile technique to study different materials. The structural changes that occur during cotton fiber development starting at 10 days post-anthesis until full maturation were successfully investigated using USTR-FTIR. The advantages of using UATRR-FTIR are that no multiple extractions are needed, no time consuming preparations of the samples are needed, ease of analysis, and the testing is non-destructive allowing other analysis to be performed on the same set of samples. Because of the potential of the FTIR as non-destructive analytical technique for cell wall components analysis, we continue developing this methodology. This could pave the way to develop a method for maturity analysis in developed cotton fibers.

REFERENCES

Abidi, N.; Hequet, E.; Cabrales, L.; Gannaway, J.; Wilkins, T.; Wells, L.W. J *Appl. Polym. Sci.* 2008, 107, 476-486.

Abidi, N. ; Hequet, E. ; Turner, C. ; Sari-Sarraf, H. *J. Applied Polymer Sci.* 2005, 96, 392-399.

Abidi, N. ; Hequet, E. *Textile Res. J.* 2007, 77(2) ,77-84.

Abidi, N.; Hequet, E. *Textile Research J.* 2005, 75(9), 645-652.

Abidi, N.; Cabrales, L.; Hequet, E. Cellulose in press, *DOI* 10.10.1007/s10570-009-9366-1.

Carpita, N.C.; Defernez, M.; Findlay, K.; Wells, B.; Shoue, D.A.; Catchpole, G.; Wilson, R.H.; McCann M.C. *Plant Physiol.* 2001, 127, 551-565.

Chen, L.; Carpita, N.C.; Reiter, W-D.; Wilson, R.H.; Jeffries, C.; McCann, M.C. *Plant J.* 1998, 16(3), 385-392.

Hequet, E.; Ethridge, D.; Cole, B.; Wyatt, B. In. Proc. 13th Annual Engineer Fiber Selection System Conference. 17-19 April 2000, Cotton Inc., Cary, NC.

Hequet, E.; Abidi N. Sticky Cotton, Measurements and Fiber Processing, Texas Tech University Press, TX, 2006ISBN 10: 0-89672-590-1.

Hendrix, D.L.; Steele, T.L.; Perkins Jr, H.H. In Bemisia: Taxonomy, biology, damage, control and management; Editor, D. Gerling and R.T. Mayer; Ed.; Intercept. UK, 1995, 189-199.

Hendrix, D.L.; Wei, Y.-A.; Leggett, J.E. *Comp. Biochem. Physiol.* 1992, 101B(1/2), 23-27.

Himmelsbach, D.S.; Akin, D.E. *Textile Research J.* 2003, 73(4), 281-288.

Hsieh, Y-L. ; Hu, X-P. ; Nguyen, A. *Textile Res J.* 1997, 67(7), 529-536.

Ilharco, L.M. ; Garcia, A.R. ; Lopez da Silva, J. ; Vieira Ferreira, L.F. Langmuir. 1997, 13, 4126-4132.

Liang, C-Y. ; Marchessault, R.H. *J. Polym. Sci.* 1959. XXXIX, 269-278.

McCann, M.C. ; Defernez, M. ; Urbanowicz, B.R.; Tewari, J.C. ; Langewisch, T. ; Olek, A.; Wells, B. ; Wilson, R.H. ; Carpita, N.C. Plant Physiol. 2007, 143, 1314-1326.

McCann, M.C.; Chen, L.; Roberts, L.; Kemsley, E.K.; Sene, C.; Carpita, N.C.; Stacey, N.J.; Wilson, R.H. *Physiol Plant.* 1997, 100, 729-738.

McCann, M.C.; Stacey, N.J.; Wilson, R.; Roberts, K. *J. Cell Sci.* 1993,106, 1347-1356.

McCann, M.C.; Hammouri, N.; Wilson, R.; Belton, P.; Roberts, K. *Plant Physiol.* 1992, 100, 1940-1947.

Miller, W.B.; Peralta, E.; Ellis, D.R.; Perkins Jr, H.H. *Textile Res. J.* 1994, 64(6), 344-350.

Morris, N.M.; Faught, S.; Catalano, E.A.; Montalvo, J.G.; Andrews, B.A.K. AATCC Review. 1994, 26(11), 33-37.

Muller, L.L.; Jacks, T.J. *J. Histochem. Cytochem.* 1975, 23(2), 107-110.

Ou-yang ,H.; Paschalis, E.P., Boskey A.L., Mendelsohn R. *Applied Spectroscopy* 2002, 56(4), 419-422.

Perkins, W.S., Textile Coloration and Finishing, Ed., Carolina Academic Press 1996.

Rajasekaran, K.; Muir, A.J.; Ingber, B.F.; French, A.D. *Textile Res. J.* 2006, 76(6), 514-518.

Sisma, S.; Schenek, A. *Milliand Textilberichte.* 1984, 13, 593-595.

Strolz, H.., In 26th International Cotton Conference, Bremen, 13-16 March, pp. 35-39, 2002.

Séné. C.F.B.; McCann, M.C.; Wilson, R.H.; Grinter, R. *Plant Physiol.* 1994, 106,1623-1631.

Watson, M.D. In. Proc. Int. Cotton Conf., 1-4 March 2000. *Faserinstitut*, Bremen, Germany.

Wei, W.; Yang, C.Q. *Textile Res. J.* 1999, 69(2), 145-151.

Wei, Y.-A.; Hendrix, D.L.; Nieman, R.J. *Agric. Food Chem.* 1997, 45, 3481-3486.

Wei, W.; Yang, C.Q. *Textile Res. J.,* 69(2), 1999, 145-151.

Wilkins, TA.; Jernstedt JA. In Cotton fibers, developmental biology, quality improvement, and textile processing, Editor, Basra A.S. Ed.; Food Products Press, NY, 1999, pp 231-269.

Yang, C.; Qian, L.; Lickfield, C. *Textile Res. J.* 2001, 71(6), 543-248.

Yang, C.; Wei, W.; Lickfield, C. *Textile Res. J.* 2000, 70(2), 143-147.

Yang, C. ; Xu, L., Shiqi L., Yangiu J., *Textile Res. J.* 1998, 68(5), 457-464.

Yong, W.; Link, B.; O'Malley, R.; Tewari, J.; Hunter, C.T. ; Lu, C-An.; Li, X.; Bleecker, A.B.; Koch, K.E., McCann, M.C.; McCarty, D.R.; Patterson, S.E.; Reiter, W-D.; Staiger, C.; Thomas, S.R.; Vermerris, W.; Carpita, N.C. *Planta*. 2005, 221, 747-751.

Zeier, J.; Schreiber, L. *Planta*. 1999, 209, 537-542.

In: Fourier Transform Infrared Spectroscopy
Editor: Oliver J. Rees, pp. 159-175

ISBN: 978-1-61668-835-6
© 2010 Nova Science Publishers, Inc.

Chapter 7

EVALUATION OF SAMPLES AND STANDARDS OF ENERGETIC MATERIALS ON SURFACES BY GRAZING ANGLE-FTIR SPECTROSCOPY

*L. C. Pacheco-Londoño, O. M. Primera-Pedrozo and S. P. Hernández-Rivera**

Center for Chemical Sensors Development
ALERT DHS Center of Excellence for Explosives
Department of Chemistry,
University of Puerto Rico-Mayagüez, Mayagüez, PR, 00681

ABSTRACT

Surface-analyte parameters, which play primary roles in the preparation of samples and standards of solid energetic materials deposited on surfaces used in the development of detection technologies for these compounds, were evaluated using Grazing Angle Probe-Fiber Optic Coupled FTIR Spectroscopy and Grazing Angle Objective FTIR micro-spectroscopy. Among the properties investigated were analyte residence time, sublimation rate from surfaces, extent and homogeneity of surface coverage, phase crystallinity, surface-analyte interactions, influence of solvent and degradation of the following high explosives: 2,4-DNT, 2,4,6-TNT, Tetryl, HTMX and RDX and the homemade energetic material TATP deposited on stainless steel surfaces. Chemometrics analysis tools, such as partial least squares, were used for quantification and determination of limits of detection and quantification. The analytical measurements were related to the parameters evaluated. Homogeneities of distributions were evaluated using point mapping analysis. Statistical figures of merit were used to classify the distributions and to interpret the results obtained. Residence times for different energetic compounds deposited on surfaces were measured at several temperatures. The variability of the properties of samples prepared was found to be dependent on residence time and surface loading homogeneity. RDX and HMX were classified as very robust candidates for sample and standard preparation based on long residence times and the quality of films deposited on the surfaces.

INTRODUCTION

The establishment of a research and development program for preparation of samples and standards is of fundamental importance in trace analysis method development and instrumentation validation programs. In the case of homeland security and national defense support, it is not only sound and proper but also required to achieve the goals aligned with defense and security applications. Proper sample development of analytes on surfaces can help achieve real, low detection limits in a remote/standoff detection program. Two aspects may be influential in the type of sample/standard prepared: whole sample detection, such as in the IMS scenario, where the complete sample is collected for analysis, or partial sample analysis, as in the case of a light beam (laser or IR) that interacts with a small part of the sample. In the latter case, a uniform distribution of analyte on the surface is required. In all cases, the most stringent requirement is that the area sampled be representative of the whole sample, both physically and chemically.

The residence time (RT) of an energetic material deposited on a surface can be described as the time the material will persist on the surface after exposure, as measured by chemical or physical analysis to a level below its low limit of detection (LOD). The RT of a solid explosive sample on a surface is affected by its sublimation rate, its vapor pressure at the substrate temperature and the nature of the interaction with surface. The term is pertinent to the development of samples used in trace detection systems (TDS) validation and testing. The sublimation enthalpy (ΔH_{sub}) is the energy change when a compound transforms from the solid phase to the gas phase. These enthalpies are present for a solid deposited on a surface. There are two general methods of obtaining the information necessary for calculation of ΔH_{sub}: direct and indirect. In the direct method of calculating this energy, a calorimeter is used. Here, the heat due to the change of phase is measured as a change in temperature in a vessel whose heat capacity contributions can be accounted for by calibration with a standard. In the indirect method of calculation, the vapor pressure, or any property proportional to it, is measured at different temperatures, and the enthalpy is calculated by use of the Clausius-Clapeyron equation [1]:

$$\ln (P/atm) = b - (\Delta H_{sub}/RT) \tag{1}$$

The homogeneity of the distribution of surface concentration (loading) of an energetic material deposited on a substrate is another aspect relevant to TDS. Many methods have been employed to study the distribution of different components in matrices and on surfaces. Spectroscopic techniques, such as Raman scattering, Near-Infrared (NIR) spectroscopy and Mid-Infrared (MIR) spectroscopy are no-destructive and no-invasive measurements, and as a result, they have been employed in numerous pharmaceutical process applications [2]-[5]. Traditional spectroscopic techniques analyze samples in bulk mode and determine an average composition across the entire sample. These conventional methods cannot determine the spatial distribution directly. In the case of samples that consist of analytes deposited on surfaces, to determine the local analyte concentration over a wide area, development of chemical imaging/mapping approaches are necessary. NIR imaging has the experimental flexibility to characterize a wide variety of samples and has been used to study the distribution of different active pharmaceutical ingredients (APIs) on relatively large areas of tablets of pharmaceutical interest [6]-[9]. Unfortunately, the spatial resolution achieved (~20-100 μm) was larger than the particle size of most ingredients. Thus, this method does not allow the study of tablets with spatial resolutions of a few micrometers or lower. Raman spectroscopy, on the other hand, can achieve a spatial resolution down to 1 to 0.5 μm [10]. However, some commonly used substances required to achieve such a high spatial resolution fluoresce with laser excitation of the sample, masking the vibrational information.

Fourier Transform Infrared (FTIR) spectroscopic imaging has emerged as an additional powerful tool to characterize distributions of chemicals in heterogeneous materials. The inherently rich vibrational information contained in the mid infrared (MIR) spectrum allows the identification of different components from their chemical structure, based on the spectral information contained in the so called "fingerprint" region, where most of the group vibrations take place ($600 - 1800$ cm^{-1}). Each FTIR spectrum represents an area of the sample as small as 10 μm, and the distribution of a single component is easily visualized by color-coding the local concentration of each species. FTIR can overcome limitations of Raman and NIR spectroscopy with proper selection conditions. The technique is much faster and sensitive. In FTIR the time domain spectrum is recorded and subsequently Fourier-transformed to the frequency domain spectrum. This way, the frequency components in a spectrum are recorded simultaneously (Fellgett's or multiplexing advantage). All the IR energy is used simultaneously, thereby achieving much higher signal-to-noise ratios (S/N) by lowering the inherent noise levels introduced by aperture slits of dispersive instruments (Jacquinot's advantage). An internal laser calibrates the interference information, providing very high wavenumber accuracy and reproducibility. As with all of modern methodologies, FTIR spectroscopy is moving into imaging/mapping technology because multichannel detectors have improved dramatically during the past decade to a point that mercury-cadmium-telluride (MCT) IR arrays can now be used for MIR spectroscopy [10]-[11].

RESIDENCE TIME AND SUBLIMATION KINETICS

Fourier Transform Infrared Spectroscopy was used to determine average residence times and to follow sublimation rates of energetic materials deposited on surfaces. The energetic materials: TATP, 2,4-DNT, TNT and RDX were deposited on stainless steel surfaces by a sample smearing method, and their sublimation was monitored using Fiber Optic Coupled Grazing Angle Probe-FTIR at several temperatures. The exponential behavior of the time decay of the IR absorbance was modeled to adjust the experimental data and diffusivity constants were determined for different temperatures for the various materials evaluated.

The samples consisted of solid energetic compounds deposited on substrates. They were prepared by transferring 20 μL aliquots of standard solutions of the compounds and placing them at one side of stainless steel (SS) plates. A Teflon sheet was inclined towards one side of the plate, and smearing was performed quickly to prevent local crystallization [12]. Methanol (MeOH), ethanol (EtOH), isopropanol (IPA; 2-propanol) and acetonitrile were used as solvents for 2,4-DNT, TNT and RDX. Dichloromethane was used for preparing TATP solutions because the high volatility of this energetic compound required a volatile solvent for dosing the analyte onto the metallic surfaces.

Surface sublimation rates were measured using a Remspec Corp. MIR grazing-angle probe (GAP) interfaced to a Bruker Vector-22 spectrometer (Bruker Optics) equipped with an external liquid nitrogen-cooled MCT infrared detector. The bench interferometer was coupled to the GAP head using a MIR fiber optic cable [12]-[14].This 1.5-m long fiber optical bundle was made of chalcogenide glass of As-Se-Te, which transmits throughout the MIR, with the exception of a strong H-Se absorbance band located about 2200 cm^{-1}. A spectral range of 1000 – 4000 wavenumbers (cm^{-1}) was used for the studies. A clean, highly reflective SS plate was used for background collection using a 4 cm^{-1} resolution. The grazing-angle head used carefully aligned mirrors to deliver the MIR beam to the sample surface at the grazing angle, approximately 80° from the surface normal. Experimental details of the setup are shown in Figure 1.

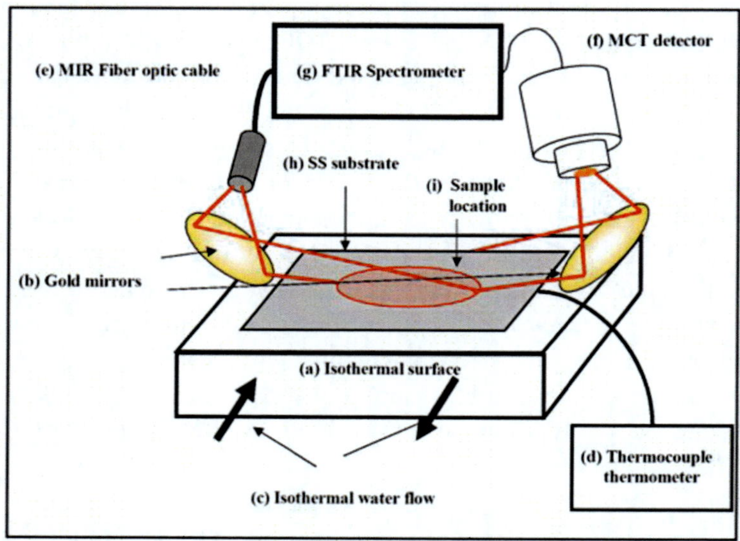

Figure 1. Schematic diagram of experimental setup for sublimation rates: (a) isothermal surface; (b) gold mirrors; (c) isothermal water flow; (d) thermocouple; (e). MIR Fiber optic cable bundle; (f) MCT detector; (g) spectrometer FTIR; (h) SS substrate; (i) sample location.

The measurements were made at controlled temperatures using a regulated water bath, which maintained a controlled temperature, with variations of ± 0.1°C. Different initial amounts of HE were deposited on the test surfaces and the variations of infrared spectra with time were recorded. Then, the temperature was increased to a new equilibrium value and spectra were recorded again. The experiments were done at different temperatures to be able to determine the relevant thermodynamic properties of the target compounds.

The MIR beam in the GAP that was directed to reflect on the test surface had an elliptical shape. The size of the ellipse was ~16 cm along the major axis and ~3 cm along the minor axis. The intensity pattern of the IR beam on the test plate had a Gaussian distribution, with an empirically determined profile described by its relative intensity (I_r) as:

$$I_r = \exp\left[-\left((0.10 \pm 0.01)x^2 + (3.3 \pm 0.2)y^2\right)\right] \qquad (2)$$

The dimensions of the ellipse that contained 95% of the total intensity were:

$$\frac{x^2}{10.0} + \frac{y^2}{0.33} = -\ln(0.05), \left(\frac{x}{5.4}\right)^2 + \left(\frac{y}{0.95}\right)^2 = 1 \quad (3)$$

The solution of Equation 2 implies that an ellipse with a major axis measuring 5.4 cm and a minor axis measuring 0.95 cm should contain 95% of the infrared beam. The corresponding ellipse necessary to capture 99.9% of the total intensity had a length of 16.4 cm on the major axis and a width of 2.9 cm on the minor axis.

Table 1 shows the experimental conditions for the energetic materials studied. For the homemade energetic compound TATP, the initial surface concentrations (Cs_o) studied were significantly higher than those of the nitroaromatic compounds (2,4-DNT and TNT) and almost two orders of magnitude higher than the nitramine (RDX). The RDX measurements were performed only at 65°C and 80°C because of the properties of the important plastic HE, such as its extremely low vapor pressure, and the experimental setup limitations.

Table 1. Calculated kinetic parameters for several energetic materials

Energetic Material	Concentration ($\mu g/cm^2$)	Temperatures (°C)
TATP	25, 50, 80	14-33
2,4-DNT	2.8, 5.7, 11.4	23-35
TNT	2.8, 5.7, 11.4	40, 60,70
RDX	0.7, 1.3	65, 70, 80

Figure 2 shows the decay of the IR signal as a function of time for the energetic material studied. For TATP, IR spectra were recorded every 12 s, and the peak area of the band located at 1330-1407 cm^{-1} was calculated for each spectrum and used in the graphical representation. This band was selected because it is well-isolated from other bands of the acetone-peroxide cyclic compound and is narrow, in comparison with the other vibrational signals of the HME. This vibration was assigned as the $\delta_{as}(CH_3)$ [15]. In the case of 2,4-DNT and TNT, both nitroaromatic compounds have strong MIR signals at 1343 cm^{-1}. This spectroscopic marker of nitroaromatic compounds was assigned to the C-NO$_2$ symmetric stretch. The vibration is coupled to the C-N stretch [1].

The band was used for monitoring the sublimation kinetics. The spectral range used for calculating peak areas was 1324-1372 cm^{-1}. For RDX a displacement in some bands was observed as RDX sublimated. The spectral range used for calculating the required peak areas was 1297-1340 cm^{-1}. The RDX band located in this range did not show appreciable displacement and is also well-isolated from other vibrational signatures. The data of peak areas (A) vs. time are shown in Figure 3 for TATP (24°C), 2,4-DNT (43°C), TNT (55°C) and RDX (80°C).

The exponential nature of the solid-vapor phase transition can be clearly observed in Figure 3. Natural logarithm fits were obtained to calculate the kinetics parameters. Results are illustrated in Figure 4.

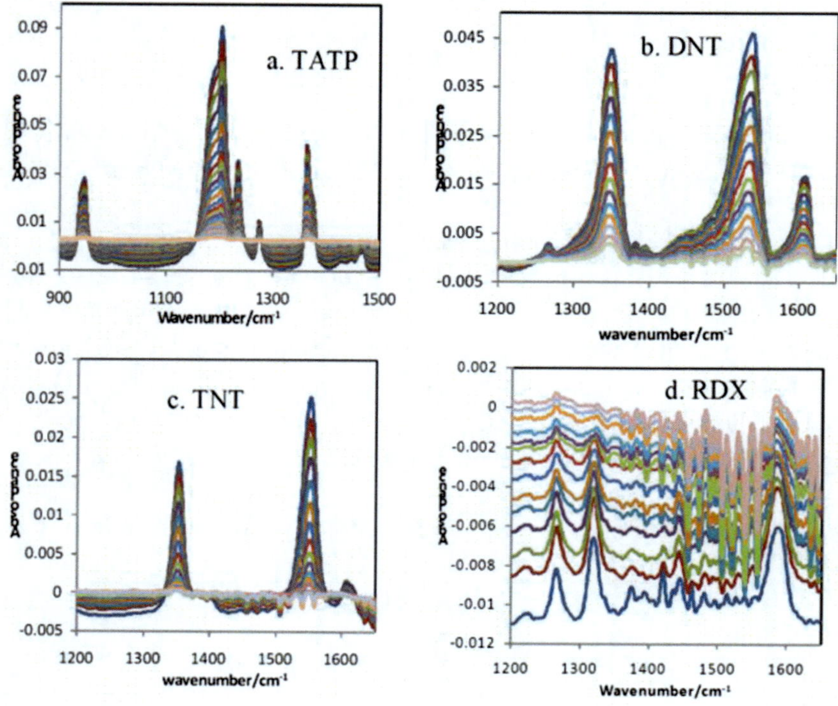

Figure 2. Time decay of IR signal (absorbance) for target chemicals: (a) 50 µg/cm² TATP at 24°C; (b) 11.4 µg/cm² 2,4-DNT at 35°C; (c) 7.6 µg/cm² TNT at 70°C; (d) 1200 ng/cm² RDX at 80°C.

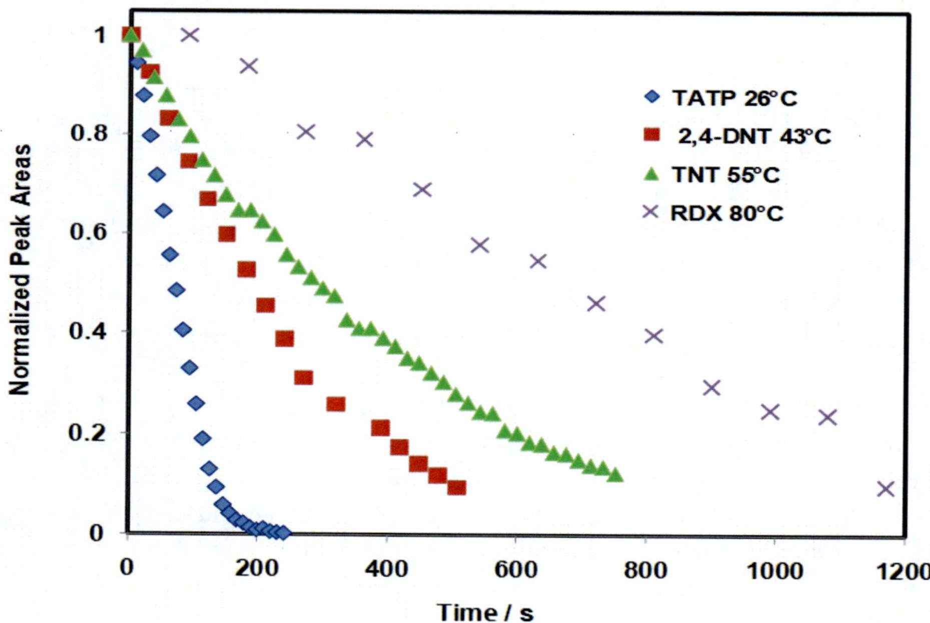

Figure 3. Sublimation rates for TATP (26°C), DNT (43°C), TNT (55°C) and RDX at (80°C).

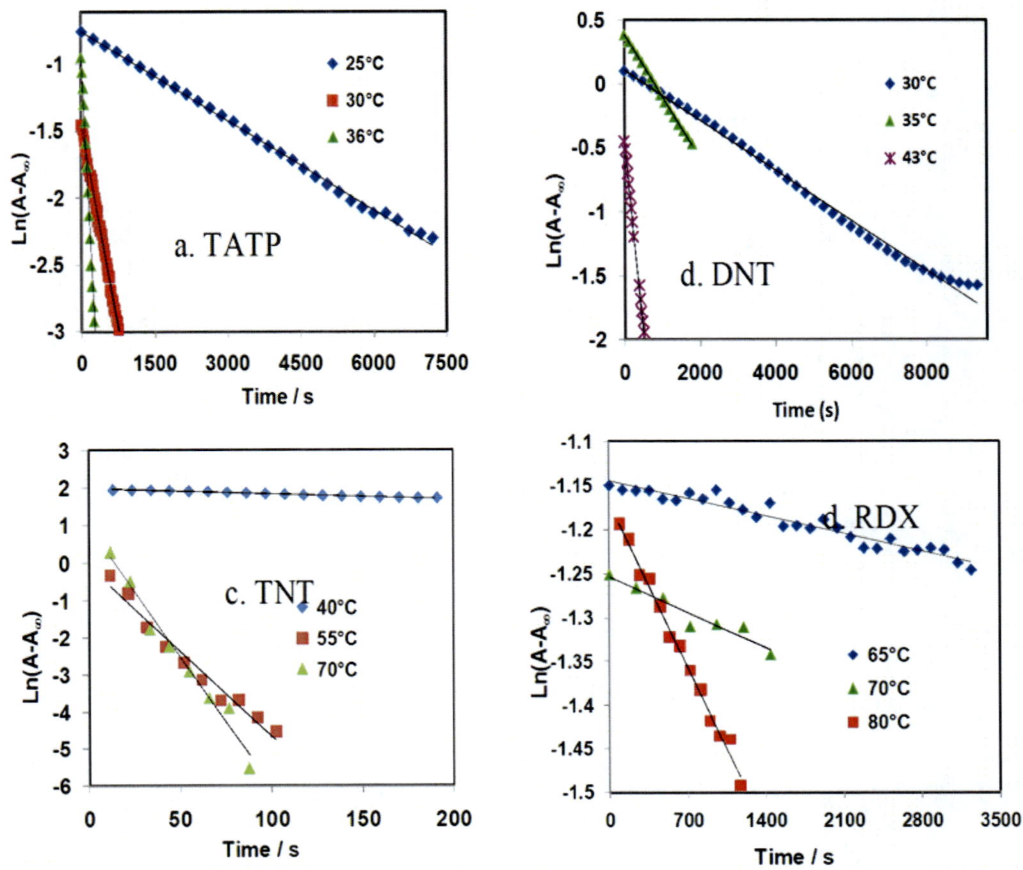

Figure 4. Ln(A-A$_\infty$) vs. time at different temperatures for: (a) TATP; (b) DNT; (c) TNT; (d) RDX.

No dependence of the kinetic parameters with the initial surface loadings (Cs$_o$) was found. A plot of ln (A-A$_\infty$) vs. time was used to obtain the sublimation rate constants (k) from the slope of the linear regressions. The fractional life for 63% of Cs$_o$ remaining on the surface (equivalent to half-life or the time required to reduce the initial concentration to the 50% value) was calculated by:

$$\ln (0.63) = - k\tau_{0.63} \tag{4}$$

The proposed model is equivalent to first-order chemical kinetics. The data is presented in Table 2. Figure 5 shows Arrhenius plots used to calculate ΔH_{sub}. The values obtained for ΔH_{sub} were 163 ± 16 kJ/mol for TATP, 100 ± 2 kJ/mol for DNT, 111 ± 7 kJ/mol for TNT and 152 ± 4 kJ/mol for RDX.

Calculated RT values as a function of temperature are shown in Figure 6. An exponential fit was performed for the different explosives, indicating that how residence time changes with the temperature. As is illustrated, the residence time for RDX at 25°C expressed as is very long.

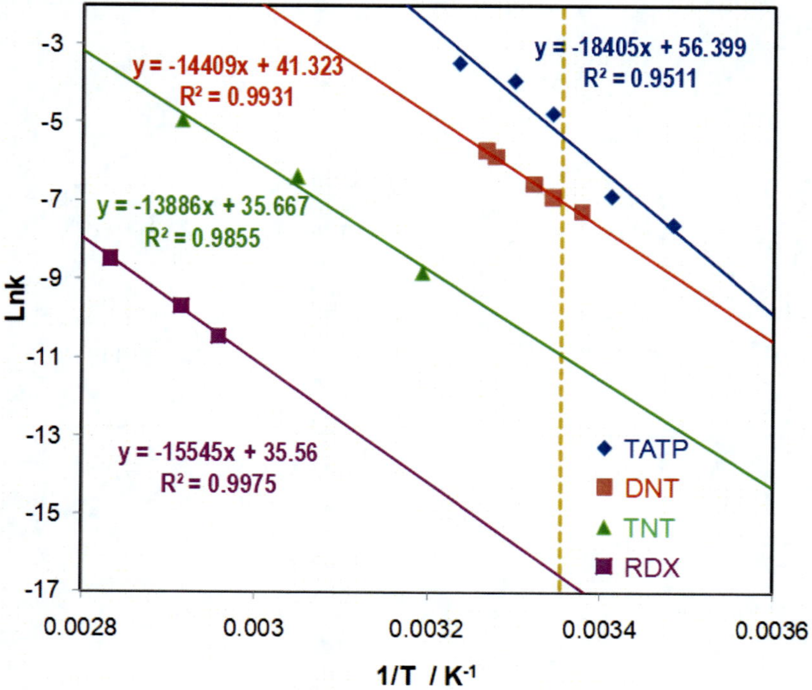

Figure 5. Arrhenius plots of sublimation rates for TATP, 2,4-DNT, TNT and RDX. The vertical line represents the reciprocal of the reference temperature (298.15 K).

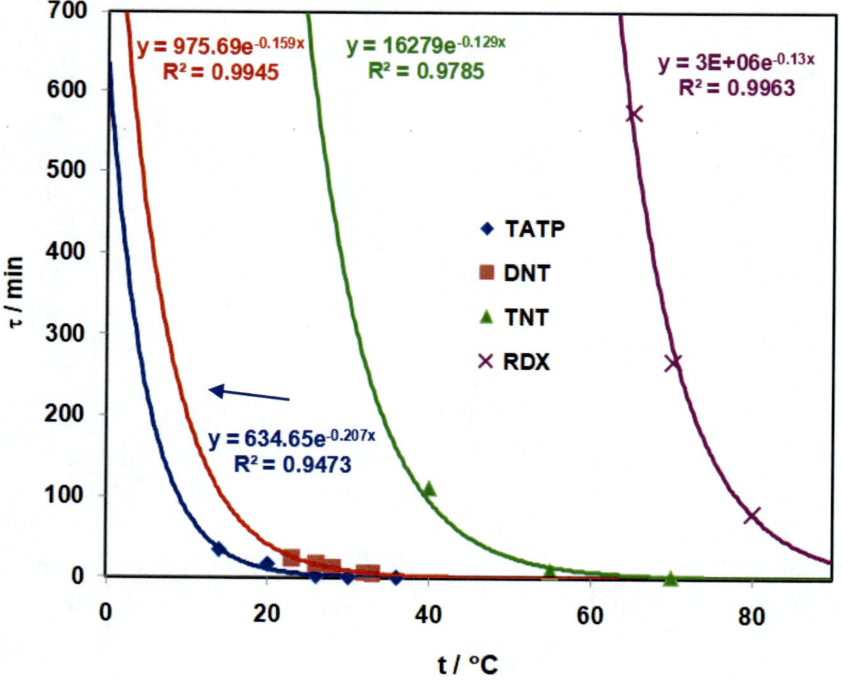

Figure 6. Residence time (RT) for TATP, DNT, TNT and RDX on SS at various temperatures.

Table 2. Calculated kinetics parameters for several energetic materials

TATP				DNT			
$t/°C$	k/s^{-1}	$1/k$	$\tau_{0.63}/min$	$t/°C$	k/s^{-1}	$1/k$	$\tau_{0.63}/min$
14	0.0005	2000	33.3	23	0.0007	1429	23.8
20	0.0010	968	16.1	26	0.0010	1000	16.7
26	0.0085	117	2.0	28	0.0014	714	11.9
30	0.0197	51	0.8	32	0.0028	357	6.0
36	0.0308	32	0.5	33	0.0033	303	5.1
TNT				RDX			
$t/°C$	k/s^{-1}	$1/k$	$\tau_{0.63}/min$	$t/°C$	k/s^{-1}	$1/k$	$\tau_{0.63}/min$
40	1.50E-04	6667	111.1	65	2.9E-05	34483	574.7
55	1.70E-03	588	9.8	70	6.2E-05	16081	268.0
70	7.15E-03	140	2.3	80	2.1E-04	4808	80.1

Extrapolated values for TNT and RDX were 24 h for TNT and ~1.3 yr for RDX. For DNT and TATP, the RT values were interpolated to 28 min for DNT and 1.9 min for TATP.

HOMOGENEITY OF DISTRIBUTIONS ON SS SURFACES

The essential parameters for describing the distributions of concentrations of energetic materials on surfaces were extracted from the images generated from vibrational maps constructed from point MIR spectra collected using a Bruker Optics IFS 66v/S FTIR interferometer, equipped with 15x magnification grazing angle objective (GAO). The images were generated from calculated peak areas for spatial grids of 150 μm x 150 μm with a separation between them of 15 μm. The intensities were color-coded to represent average ranges of peak areas. Typical vibrational maps for TNT on SS surfaces deposited using methanol (MeOH), ethanol (EtOH), acetonitrile, acetone and isopropanol (IPA) are shown in Figure 7. Vibrational maps for HMX, TNB and Tetryl using isopropanol as a solvent are also shown in Figure 7 [16].

Calibration curves were obtained using the PLS1 regression algorithm from the chemometrics routine [17]-[19] Quant2 software of Bruker Optics OPUS (version 4.2) [20]. The 36 standards used for each of the calibration curves were prepared by sample smearing at room temperature. Using the calibration curves obtained for each compound, peak areas were then converted to surface loadings [21]. Next, histograms of the maps were generated. Statistical parameters used to characterize the distributions, such as the standard deviations, are among the important descriptors utilized in describing the distributions obtained. Standard deviations (SD) are good measures of the degree of homogeneity of the distributions of HE on the test surfaces. Other parameters used were: mean values, kurtosis (attributes of distributions describing the "peakedness" and whose expected value is 3), and skewness (a distribution that is symmetric about its mean has skewness of zero) and the moments of the distributions. These parameters are customarily calculated in a distribution and are used for measuring deviation from an otherwise normal (Gaussian) distribution.

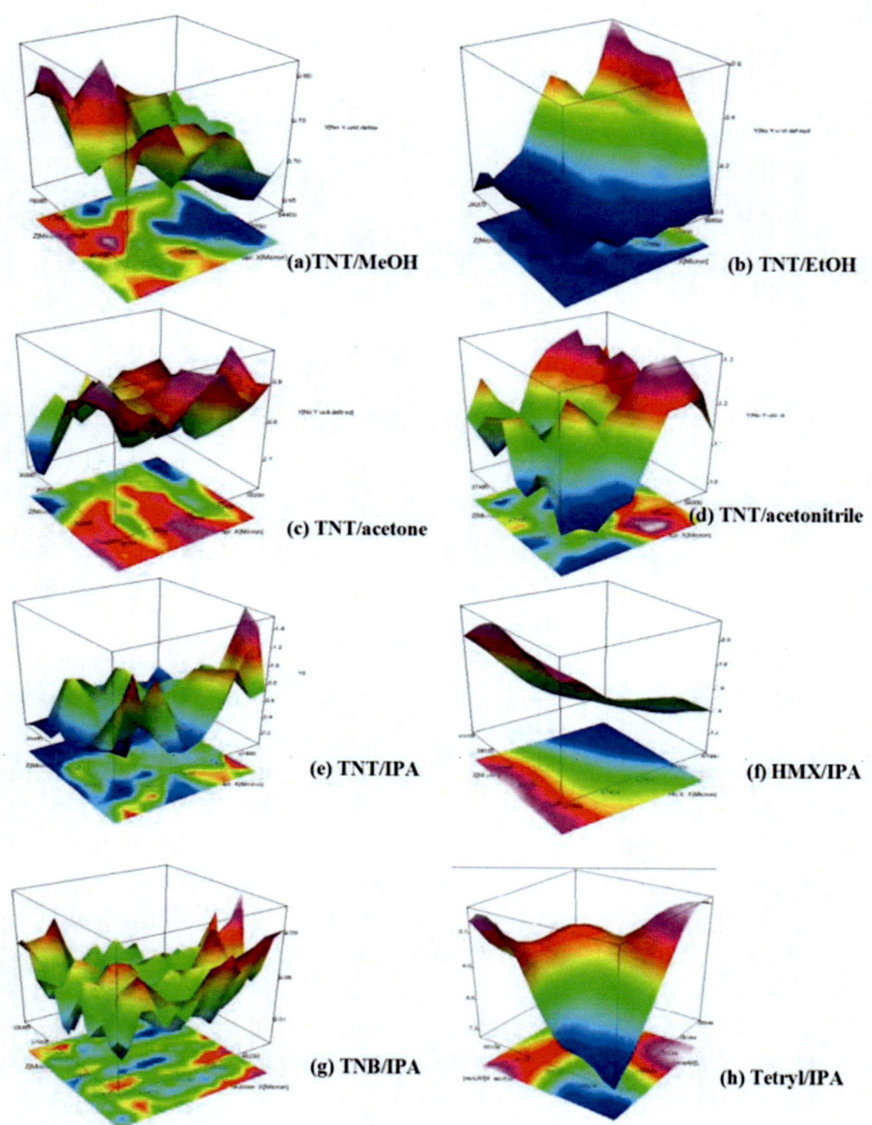

Figure 7. Maps of distributions on SS substrates: (a) TNT/methanol; (b) TNT/ethanol; (c) TNT/acetone; (d) TNT/acetonitrile; (e) TNT/IPA; (f) HMX/IPA; (g) TNB/IPA; (h) Tetryl/IPA.

Nominal surface concentrations of: 10.0 μg/cm^2 of TNT, 4.8 μg/cm^2 of HMX, 7.0 μg/cm^2 of TNB and 8.1 μg/cm^2 of Tetryl are illustrated. The contour for the deposition of TNB with isopropyl alcohol as the solvent has a random appearance. In the HE-solvent combinations of TNT/MeOH, TNT/acetone, TNTT/acetonitrile and TNT/IPA, the solute/solvent pairs also have random appearances. This was demonstrated by an analysis of the frequency distribution of the surface concentration data, as assessed in as many points as is statistically significant. If the distribution is close to the normal distribution (Gaussian), this would indicate the presence of a random distribution. This analysis was done for all maps of surface loadings.

Table 3 lists a summary of all statistical measures for each explosive deposited on SS surfaces. It includes measures of central tendency, measures of variability and measures of shape. Of particular interest here are the standardized skewness and standardized kurtosis, which can be used to determine whether the sample distribution on the substrate behaves as a normal distribution. Values of these statistics outside the range of -2 to +2 indicate significant departures from "normality", which would tend to invalidate many of the statistical procedures normally applied to this data. Distributions obtained using ethanol and isopropanol as solvent show standardized skewness values outside the expected range. Tetryl and IPA solute/solvent combinations show standardized kurtosis values outside the expected range.

Table 3. Statistical parameters for the distributions of He on SS substrates

Para-meter	TNT/ MeOH	TNT/ ace-tone	TNT/ acetone-trile	TNT EtOH	TNT/ IPA	HMX /IPA	TNB/ IPA	Tetryl/ IPA
Count	100	100	100	100	100	100	100	100
Average	0.72	0.82	1.15	0.5	0.5	1.6	0.080	0.86
Std. dev.	0.04	0.08	0.08	0.2	0.3	0.2	0.005	0.10
Coeff. variation	5.9%	10.2 %	7.1%	28.2%	56.0%	14.3 %	6.6%	11.3%
Minimum	0.64	0.62	0.98	0.35	0.08	1.16	0.07	0.67
Maximum	0.82	0.98	1.32	0.96	1.53	2.07	0.10	1.06
Range	0.18	0.35	0.34	0.61	1.45	0.91	0.03	0.39
Std. skewness	0.43	-1.67	-0.13	3.45	4.42	0.96	1.93	0.54
Std. kurtosis	-1.31	-1.16	-1.62	-0.029	2.90	-1.43	-0.42	-2.14

In the cases of depositions of HMX and TNB with 2-propanol (IPA) as solvent, homogeneous distributions were observed. This phenomenon was not observed for other solvents (data no shown). The smearing deposition method performed very well in depositing the energetic materials on the surfaces investigated, producing normal distributions or near to normal distribution of analytes on the test surface. The errors or variations in the infrared absorbance (areas) for different samples of a given nominal surface loading were smaller than the distribution uncertainties. Other factors may be responsible for these variations.

For the case of RDX, improved film-like, homogeneity was observed using an unsaturated solution in IPA. A very low coefficient of variation (CV), in comparison with other target compounds, was found for a Cs_0 of 500 ng/cm^2 RDX/IPA (Table 4). The skewness and kurtosis were close to one, indicating a narrow distribution. However, CV values increased with higher initial surface loadings (Cs_o) because a more concentrated solution of RDX in IPA was necessary and led to near to saturation in isopropanol. At this point, a film was not observed, but rather droplets of a metastable phase were obtained.

Table 4. Statistical parameters for the distributions of RDX in isopropanol on SS

Parameter		Value
Count		210
Average		3.96
Std. deviation		0.03
Coeff. of variation (CV)		0.63%
Minimum		3.90
Maximum		4.02
Range		0.123
Std. skewness		0.94
Std. kurtosis		-1.07

METASTABLE PHASES OF TNT AND RDX

The formation of metastable phases on stainless steel substrates has been well established for TNT [22]-[25] and RDX [26]. In some cases, the formation of metastable aggregation of particles was a highly persistent phenomenon, at times nearly unavoidable, for certain energetic material/solvent/deposition method combinations. Figure 8 shows white light micrographs of the metastable phase found for RDX obtained from solvent evaporation and agglomeration of small particles. Figure 8-a shows a 50x magnification micrograph; Figure 8-b shows 100x magnification of the metastable phase of RDX formed by solvent evaporation and agglomeration. Deposition of RDX was performed using highly dilute solutions of analyte in a highly volatile solvent. Metastable RDX phases from deposition of RDX/IPA using saturated solutions are also shown: Figure 8-c: 50x magnification; Figure 8-d: 100x magnification. Sizes of images shown were: 50x: 114 μm x 85 μm; 100x: 57μm x 43 μm.

A mechanism for the transition from metastable phases to crystalline phases was proposed. The metastable phase of TNT can be converted to the room temperature stable crystalline phase by friction or by application of slight pressure [26]. The metastable phase of RDX (m-RDX) was obtained when the sample was allowed to sublimate and condense on a glass slide (method 1) or by depositing the sample from dilute solutions (acetone, methanol or saturated in isopropanol) on SS as shown in Figure 8.

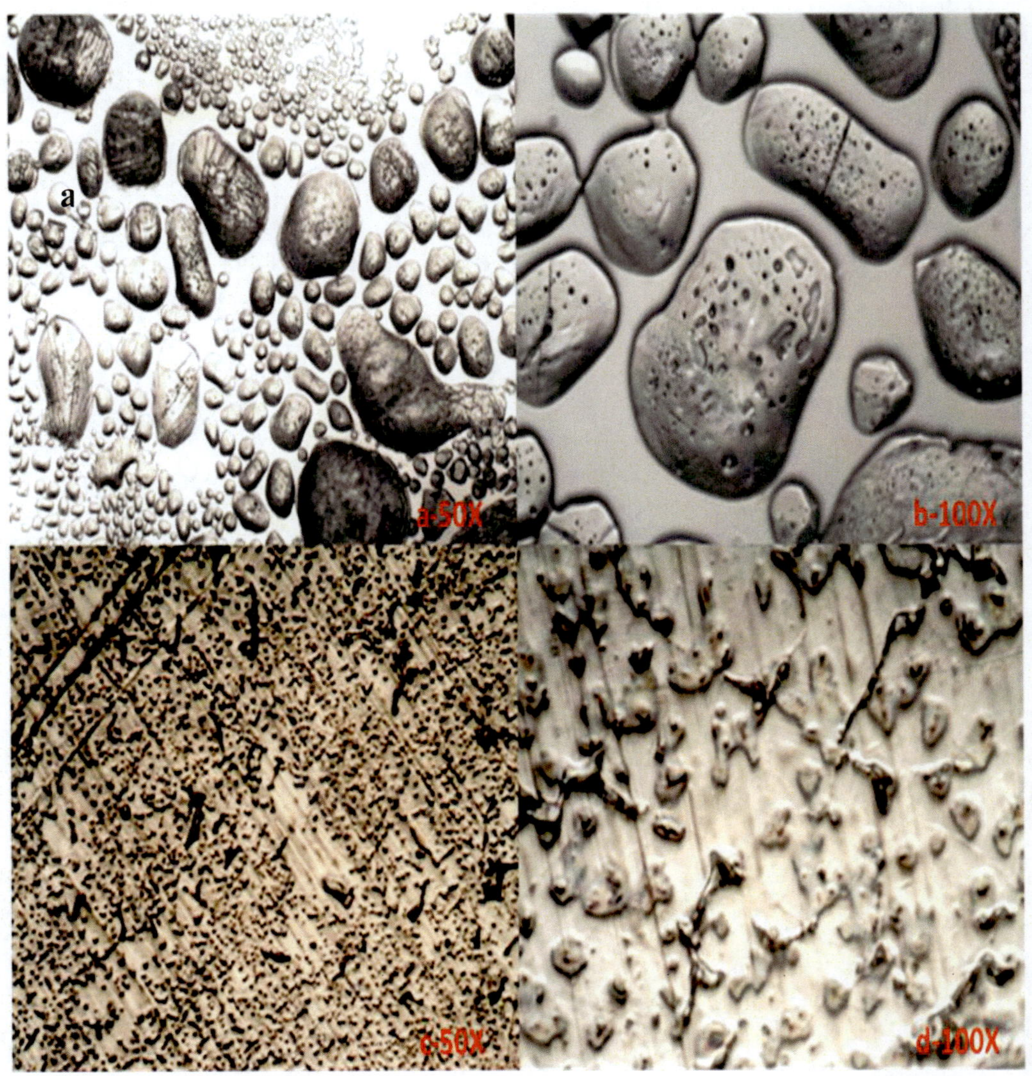

Figure 8. M-RDX phase obtained from solvent evaporation and agglomeration: (a) 50x; magnification; (b) 100x magnification. M-RDX phase from deposition of saturated solution of RDX in IPA: (c) 50x magnification; (d) 100x magnification. Sizes: 50x: 114μm x 85 μm; 100x: 57μm x 43 μm.

The RDX conformation in the metastable phase found was spectroscopically similar to the α-RDX phase [26]-[28]. This was verified by comparison of the macro grazing-angle (60° from surface normal, GAIR-60°) mid infrared spectrum of m-RDX on SS and the transmittance spectrum of solid RDX mixture with KBr, as shown in Figure 9. Bulk RDX spectra for the room temperature stable solid exhibits two peaks at 1574 cm^{-1} and 1596 cm^{-1}. Similarly, m-RDX GAIR-60° shows the two bands assigned to asymmetric N-O stretching [29]. The C-H stretching region was almost identical for the two spectra, and the region of 400 – 1800 was also nearly identical.

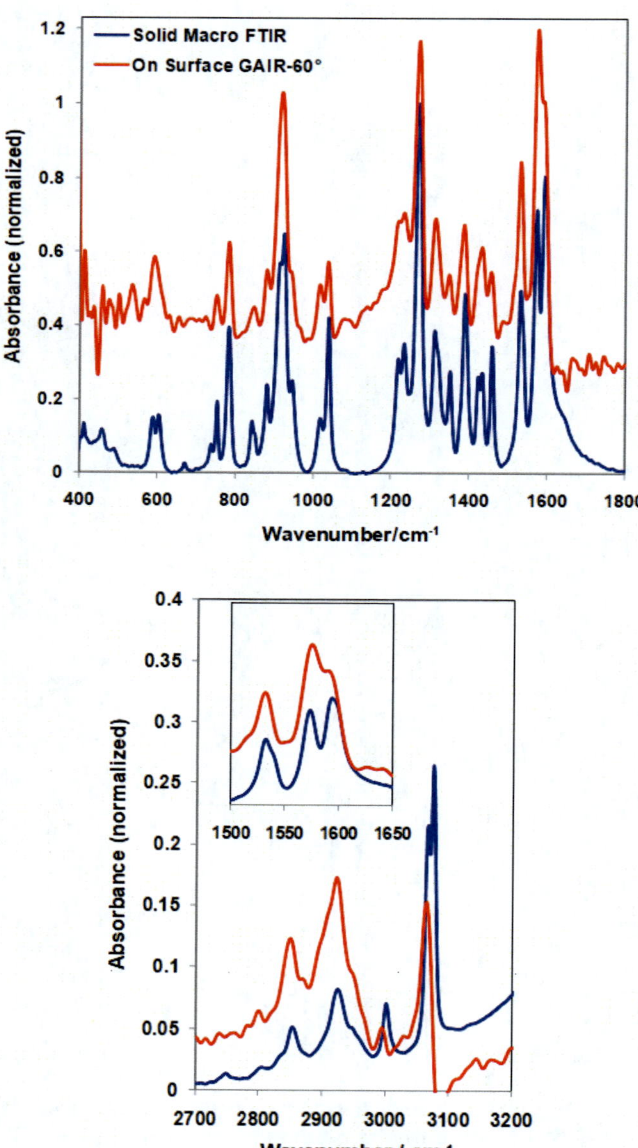

Figure 9. MIR grazing-angle spectrum at 60° from surface normal (GIR 60°) of m-RDX on SS and comparison with spectrum of solid by transmittance: (a) fingerprint region; (b) CH region.

CONCLUSION

The rates of sublimation of four energetic materials deposited on stainless steel substrates were studied using grazing angle probe-fiber optics coupled to FTIR operating in the mid infrared region. The technique was successful in accomplishing the goals because it accounts for the total mass of the analytes deposited on the surface. The rate of decay of the MIR signal was used to determine both sublimation kinetics and residence times of the selected compounds on surfaces coated with target compounds. Residence times on the stainless steel substrates were extrapolated for TNT and RDX and interpolated for DNT and TATP. The values obtained for τ were ~1.3 year for RDX, 24 hours for TNT, 28 minutes for DNT and 1.9 minutes for TATP. The values obtained for τ were for the energetic materials on the test surface. These values may be different for macro particles of energetic materials because the absence of analyte/surface interactions in macro (bulk) particles. These differences can also affect the enthalpies obtained.

The smearing deposition method, used for preparing samples and standards of energetic materials loaded on surfaces, worked successfully for depositing the compounds on the surfaces investigated when using dilute concentrations of analytes in appropriate solvents. Normal or near normal distributions of analytes on the test surfaces were obtained. The errors or variations in the infrared absorbencies (peak areas) for different samples of a given nominal surface loading are smaller than the uncertainties in the distribution statistical parameters. Solvent vapor pressure plays a primary role in determining the morphology of the deposits of energetic materials on the substrates. Highly volatile solvents, such as methanol, tend to stimulate the formation of thin layers of solid of the compounds tested. On the contrary, solvents that stay on the surface longer promote the formation of islets, giving rise to highly non-uniform distributions of analyte on the substrates. RDX deposited from a non saturated solution in isopropanol also showed good homogeneity distributions.

The formation of crystalline islets vs. formation of metastable phases of TNT and RDX was also investigated. A metastable phase for RDX could be obtained by two deposition methods. The formation/selection of crystalline/metastable phases opens new pathways for investigation of HE on surfaces.

ACKNOWLEDGMENTS

Parts of the work presented in this contribution were supported by the U.S. Department of Defense, University Research Initiative Multidisciplinary University Research Initiative (URI)-MURI Program, under grant number DAAD19-02-1-0257. The authors also acknowledge contributions from Aaron LaPointe of Night Vision and Electronic Sensors Directorate, Department of Defense, Jennifer Becker MURI Program Manager, Army Research Office and Stephen J. Lee Chief Scientist, Science and Technology, Office of the Director, Army Research Office/Army Research laboratory

Support from the U.S. Department of Homeland Security under Award Number 2008-ST-061-ED0001 is also acknowledged. However, the views and conclusions contained in this document are those of the authors and should not be interpreted as necessarily representing the official policies, either expressed or implied, of the U.S. Department of Homeland Security.

REFERENCES

[1]   Levine, I. N. Physical Chemistry. 7th ed. New York, NY: McGraw-Hill Higher Education. 2009. pp. 213-217.San Diego, CA: Academic Press.1991.

[2]   Mehta, N. K.; Goenaga-Polo, J. E.; Hernández-Rivera, S. P.; Hernández, D.; Thomson, M. A; Melling, P. J. *Bio Pharm.* 2002, 15, 36-42.

[3]   Primera-Pedrozo, O. M. Fiber Optic Coupled Grazing Angle/Fourier Transform Reflection Absorption Infrared Spectroscopy as an Analyzer of Explosive and Pharmaceutical Residues on Surfaces. University of Puerto Rico, Mayaguez Campus, Mayaguez, PR: 2005.

[4]   Melling, P. J.; Shelley, P. "Spectroscopic Accessory for Examining Films and Coatings on Solid Surfaces" U.S Patent 6,3,10,348, United States Patent and Trademark Office, Washington, DC.

[5]   Hamilton, M. L.; Perston, B. B.; Harland, P. W.; Williamson, B. E.; Thomson, M. A.; Melling, P. J. Grazing-Angle Fiber-Optic IRRAS for *in situ* Cleaning Validation. *Org. Process Res. Dev.* 2005, 9, (3), 337-343.

[6]   Ellison, C. D.; Ennis, B. J.; Hamad, M. L.; Lyon, R. C. Measuring the distribution of density and tabletting force in pharmaceutical tablets by chemical imaging. *Journal of Pharmaceutical and Biomedical Analysis* 2008, 48, (1), 1-7.

[7]   Reich, G. Near-infrared spectroscopy and imaging: Basic principles and pharmaceutical applications. *Advanced Drug Delivery Reviews* 2005, 57, (8), 1109-1143.

[8]   Clarke, F. C.; Jamieson, M. J.; Clark, D. A.; Hammond, S. V.; Jee, R. D.; Moffat, A. C. Chemical Image Fusion; The Synergy of FT-NIR and Raman Mapping Microscopy To Enable a More Complete Visualization of Pharmaceutical Formulations. *Analytical Chemistry* 2001, 73, (10), 2213-2220.

[9]   Sekulic, S. S.; Ward, H. W.; Brannegan, D. R.; Stanley, E. D.; Evans, C. L.; Sciavolino, S. T.; Hailey, P. A.; Aldridge, P. K. On-Line Monitoring of Powder Blend Homogeneity by Near-Infrared Spectroscopy. *Analytical Chemistry* 1996, 68, (3), 509-513.

[10]   Keen, I.; Rintoul, L.; Fredericks, P. M. Raman and Infrared Microspectroscopic Mapping of Plasma-Treated and Grafted Polymer Surfaces. *Appl. Spectrosc.* 2001, 55, (8), 984-991.

[11]   Schultz, C. P. *Precision Infrared Spectroscopic Imaging: The Future of FT-IR Spectroscopy.* Advanstar Communications, Inc., Woodland Hills, CA: October 1, 2001.

[12]   Primera-Pedrozo, O. M.; Pacheco-Londoño, L.; Ruiz, O.; Ramirez, M.; Soto-Feliciano, Y. M.; De La Torre-Quintana, L. F.; Hernandez-Rivera, S. P. Characterization of thermal inkjet technology TNT deposits by fiber optic-grazing angle probe FTIR spectroscopy. *Proc. SPIE* 2005, 5778, 543-552.

[13]   Primera-Pedrozo, O. M.; Pacheco-Londoño, L. C.; De la Torre-Quintana, L. F.; Hernandez-Rivera, S. P.; Chamberlain, R. T.; Lareau, R. T. Use of fiber optic coupled FT-IR in detection of explosives on surfaces. *Proc. SPIE* 2004, 5403, 237-245.

[14]   Buttigieg, G. A.; Knight, A. K.; Denson, S.; Pommier, C.; Denton, M. B. Characterization of the explosive triacetone triperoxide and detection by ion mobility spectrometry. *Forensic Science International* 2003, 135, 53-59.

[15]   Clarkson, J.; Smith, W. E.; Batchelder, D. N.; Smith, D. A.; Coats, A. M. A theoretical study of the structure and vibrations of 2,4,6-trinitrotoluene. *Journal of Molecular Structure* 2003, 648, 203-214.

[16]   Primera-Pedrozo, O. M.; Soto-Feliciano, Y. M.; Pacheco-Londono, L. C.; Hernandez-Rivera, S. P. Detection of High Explosives Using Reflection Absorption Infrared

Spectroscopy with Fiber Coupled Grazing Angle Probe/FTIR. *Sens Imaging* 2009, 10, 1-13.

[17] Otto, M. Chemometrics: Statistics and Computer Application in Analytical Chemistry. Weinheim: Wiley-VCH. 1999.

[18] Brereton, R. G. Chemometrics: Data Analysis for the Laboratory and Chemical Plant. John Wiley and Sons. 2003.

[19] Beebe, K. R.; Pell, R. J.; Seasholtz, M. B. Chemometrics: A Practical Guide. John Wiley and Sons. 1998.

[20] OPUS 4.2 Version. User Manual. Bruker Optics: Billerica, MA. 2003.

[21] IUPAC. Commission on Spectrochemical and Other Optical Procedures for Analysis: Nomenclature, Symbols, Units and Their Usage in Spectrochemical Analysis-II. Data Interpretation. *Pure Appl. Chem* 1976. 45, 99.

[22] Manrique-Bastidas, C.; Castillo-Chará, J.; Mina, N.; Castro, M.; Hernández-Rivera, S. P. Nucleation and crystallization studies: a vibrational spectroscopy investigation of 2,4,6-TNT. *Proc. SPIE-Int. Soc. Opt. Eng.* 2004, 5415, 1345.

[23] Vrcelj, R. M.; Gallagher, H. G.; Sherwood, J. N. Polymorphism in 2,4,6-Trinitrotoluene Crystallized from Solution. *Journal of the American Chemical Society* 2001, 123, (10), 2291-2295.

[24] Vrcelj, R. M.; Sherwood, J. N.; Kennedy, A. R.; Gallagher, H. G.; Gelbrich, T., Polymorphism in 2-4-6 Trinitrotoluene. *Crystal Growth and Design* 2003, 3, (6), 1027-1032.

[25] Manrique-Bastidas, C.; Primera-Pedrozo, O.; Pacheco-Londoño, L.; Hernández-Rivera, S., Raman microspectroscopy crystallization studies of 2,4,6-TNT in different solvents. *Proc. SPIE-Int. Soc. Opt. Eng.* 2004, 5617, 429.

[26] Pacheco-Londoño, L. C.; Ortiz-Rivera, W.; Primera-Pedrozo, O. M.; Hernández-Rivera, S. P. Vibrational spectroscopy standoff detection of explosives. *Anal. Bioanal. Chem.* 2009, 395, 323–335.

[27] Torres, P.; Mercado, L.; Cotte, I.; Hernandez, S. P.; Mina, N.; Santana, A.; Chamberlain, R. T.; Lareau, R.; Castro, M. E. Vibrational Spectroscopy Study of b and a RDX Deposits. *J. Phys. Chem. B* 2004, 108, 8799-8805.

[28] Kuklija, M. M.; Kunz, A. B. Simulation of Defects in Energetic Materials. 3. The Structure and Preoperties of RDX Crystals with Vacancy Complexes. *J. Phys. Chem. B* 1999, 103, 8427-8431.

[29] Karpowicz, R. J.; Brill, T. B. Comparison of the Molecular Structure of Hexahydro-I ,3,5-trinitro-s-triazine in the Vapor, Solution, and Solid Phases. *The Journal of Physical Chemistry* 1984, 88, 348-351.

In: Fourier Transform Infrared Spectroscopy   ISBN: 978-1-61668-835-6
Editor: Oliver J. Rees, pp. 177-194   © 2010 Nova Science Publishers, Inc.

Chapter 8

AN *IN SITU* FTIR FIBER OPTIC METHOD FOR THE DETECTION OF ACTIVE PHARMACEUTICAL INGREDIENTS AND EXCIPIENTS ON METALLIC SUBSTRATES

P. M. Fierro-Mercado, O. M. Primera-Pedrozo, A. Hornedo and S. P. Hernández-Rivera

Center for Chemical Sensors Development
Chemical Imaging Center
Department of Chemistry, University of Puerto Rico-Mayagüez
Mayagüez, PR 00681-9000
Bristol-Myers-Squibb
PR State Rd. # 3 Km. 77.5 Humacao,
PR 00791-0609

ABSTRACT

A simple, rapid and low-cost method is presented as a new alternative for the detection of trace amounts of organic compounds on surfaces. This methodology uses optical fibers coupled to a Grazing Angle Probe-Fourier Transform Infrared Spectrometer, which allows remote sensing and direct detection of contaminants left on the surfaces of pharmaceutical reactors. This method is useful for modern programs in cleaning validation, is solvent free and requires no sample preparation. Smearing deposition was used to transfer the target analyte on the substrates to be used as samples and standards. Samples of an active pharmaceutical ingredient, provided by Bristol-Myers Squibb in Humacao Puerto Rico, and magnesium stearate, used as an excipient in concentrations ranging from 0.07 to 10.0 $\mu g/cm^2$, were deposited on stainless steel metal surfaces. Methanol was used as the transfer solvent for smearing. The amount of analyte was related to the intensity of absorption bands due to fundamental vibrations of the analyte in the fingerprint region of the mid-infrared spectra. To establish the relationship between the deposited amount of analyte and the intensity of the infrared signals, a multivariate calibration procedure using partial least squares regression and a discriminant analysis coupled with principal component analysis was used. The proposed method has a limit of detection 280 ng/cm^2 with a relative error of 3.6%.

INTRODUCTION

Technical developments in fiber optic technology encouraged us to evaluate a mid-infrared (MIR) fiber optic-based sensing scheme for the detection of organic traces, which demonstrated great potential in industrial applications such as cleaning validation processes [1]-[2]. Cleaning validation is the process of ensuring that cleaning practices effectively eliminate the residues from manufacturing equipment or facilities to below a predetermined level, especially from surfaces in process equipment prior to its use for another purpose, such as batch changeover. [3]. This is particularly important in the pharmaceutical industry when equipment is used for processing two or more active pharmaceutical ingredients (APIs) and where cross-contamination could have severe consequences. An ideal validation method for cleaning procedures would be a rapid, automated, *in situ,* solvent-less and multi-component analysis of the entire surface [4].

A recently developed technique that combines a mid-infrared (MIR) interferometer with a grazing angle incidence reflection accessory has the ability to detect low chemical concentrations on reflective surfaces such as metals and non reflective surfaces, including plastic and glass [5]-[6]. Fiber optic materials that are transparent in the MIR spectral region offer access to the fundamental vibrational fingerprint absorption of organic molecules, which imprints inherent molecular selectivity.

In the last decade, MIR fiber optic probes have gained acceptance as convenient and useful tools for chemical sensing in a wide variety of applications [7]. Fiber optics has made possible many measurements that would otherwise be impractical. They have impacted nearly all areas of science and technology from the medical field to communications and have matured to the point that they are routinely used in laboratory environments by non-spectroscopists. Infrared fibers provide the ability to take the spectrometer to the sample instead of the traditional method of taking the sample to the spectrometer.

An MIR fiber-optic coupled grazing angle probe (GAP) that measures specular reflection has been developed for analysis of trace residues deposited on surfaces as contaminants. The GAP is comprised of a 19-fiber cable, which brings the signal from the spectrometer to the probe head. It is compact, fast and ready for *in situ* analysis outside of the sample compartment of a bench MIR spectrometer. Other attractive features offered by this technique include the following: portability, ease of use, rugged design, high sensitivity and short analysis time. This novel setup can be used to detect and quantify small amounts of organic materials left on surfaces [8]. Coupling of the spectroscopic hardware with powerful statistical chemometrics routines has led to a powerful technique for surface contamination detection and measurement [9].

Grazing-angle Probe Fourier Transform Infrared spectroscopy with a partial least squares model (PLS) can improve the current technique for determining the presence of organic matter on surfaces. This procedure appears to be a very promising method for rapid and low-cost determination of residues left on the surfaces of pharmaceutical reactors; however, it can also be used in surface corrosion studies, for detecting traces of environmental contaminants on surfaces and in detection, identification and quantification studies of high explosives and mixtures on surfaces [10]-[12].

BACKGROUND AND RATIONALE

An important step in pharmaceutical processing consists of the removal of drug residues from the equipment and areas involved in the various manufacturing stages. Industries face a significant problem of validating the cleanliness of surfaces that come in contact with active pharmaceutical ingredients (APIs) because a large variety of potent materials are used to manufacture a wide spectrum of different products. The procedure used to clean the drug residues must be validated according to the good manufacturing practices (GMP) rules and guidelines; therefore, special attention is required when selecting the methods used to determine trace amounts of drugs.

Cleaning validation of surfaces that come in contact with APIs can be complex and time-consuming processes because residues can be left attached to the surface by a combination of electrostatic forces and mechanical adhesion. Health-related industries utilize various approaches for cleaning validation, including swab sample collection, solvent extraction and HPLC analysis. These approaches are expensive, time consuming and inconsistent due to the non-uniform distribution of contaminants left as minute residues of API. Moreover, there are always possibilities of introducing further, indeterminate sources of contamination due to the handling and treatment of samples for analytical procedures.

This chapter focuses on the utilization of fiber optic coupled-Fourier Transform infrared spectroscopy for the detection of organic traces for cleaning validation processes without collecting the sample. A solvent-free, remotely-sensing, spectroscopic technique that uses a Mid-IR fiber optic grazing probe has been developed to measure low amounts of residues of chemicals left on surfaces as contaminants. The method is environmentally safe and fast, and it eliminates the possibility of contaminating the sample to be analyzed due to handling and pretreatment.

SAMPLE PREPARATION

Appropriate methods for standard preparation include smearing [2], spraying [9] and thermal inkjet (TIJ) deposition [10]. Spray and TIJ methodologies require an independent method to establish the exact amount of the target compound deposited. The smearing methodology is rapid, simple and easy to execute. The amount of applied material can be calculated without an independent analysis; however, it can be cross-checked for incomplete transfer using a primary method of analysis, such as gas chromatography or high performance liquid chromatography (HPLC). In summary, a volume of sample of known concentration is placed at one side of the plate and smeared over the surfaces using a Teflon sheet, inclined towards the right or left, in a single pass operation while drops of solvent disappear. This operation can be repeated back and forth until the distribution of analyte on the surface is homogeneous.

In order to assess the degree of sample transfer, external and internal validations were performed. The results were compared with those obtained using a primary method of analysis based on HPLC as a reference method used in pharmaceutical manufacturing. The metal plates were washed using 5-10 mL of methanol to completely remove the applied target compound, while the surface loading was back-calculated using HPLC according to method described by Erk [11].

IN SITU FTIR-BASED SCREENING TECHNIQUE

An IR spectrometer, when used with a grazing angle incidence reflection accessory, can detect and quantify low chemical concentrations on surfaces such as metals at substrate loadings below 10 $\mu g/cm^2$, making it applicable for cleaning validation. A detection, identification, quantification and discrimination methodology was developed based on the interfacing of an MIR interferometer with a reflection-absorption accessory. A high sensitivity Grazing Angle Probe (GAP) head and coupled to chemometrics-based, statistical routines for chemical analyses is capable of screening a surface in seconds and detecting low to very low concentration levels of surface contaminants. The feasibility of this system has been demonstrated for various substrates, such as stainless steel, plastic and glass, and has been previously described in detail [5], [10]-[12].

MIR spectroscopy operating at the grazing angle of incidence is one of the most sensitive optical absorption techniques available for measuring low chemical concentrations on reflective surfaces such as metals. A grazing angle specular fiber-optic reflection probe consists of a 19-fiber cable that is used to bring the signal from the spectrometer to the probe head, and either another 19-fiber cable returns the signal to a remote detector, or a detector is mounted directly on the probe, as illustrated in Figure 1.

In this probe, the signal exiting the fiber cable is collimated using an off-axis parabolic gold mirror and is directed towards the sample at 80° from normal. It is then refocused by another off-axis parabolic gold mirror into the return fiber cable or a detector element, in this case, an external mercury-cadmium-telluride (MCT) detector. Typically, MIR grazing-angle spectra or Infrared Reflection Adsorption Spectroscopy (IRRAS) spectra are recorded by averaging 50 scans with a resolution of 4 cm^{-1} over the range of 4000 – 1000 wavenumbers (cm^{-1}) using the Bruker Optics OPUS™ software package for data collection.

IRRAS is a single beam technique, for which a background spectrum is scanned before each measurement session using a clean test metal plate at the same instrumental conditions used for sample spectrum acquisition. All spectra were recorded in absorbance mode. The calibration models are built using the QUANT 2 chemometrics package of OPUS BRUKER software, which is based on the PLS1 algorithm. Figure 2 shows the transmission single beam spectrum of the chalcogenide glass fiber optic bundle used in the GAP, which transmits throughout the mid-IR with the exception of a strong H-Se absorbance band at 2200 cm^{-1}.

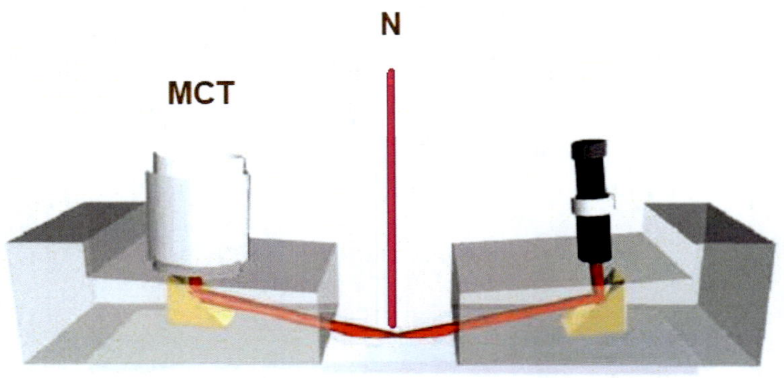

Figure 1. Grazing angle probe (GAP): sensor head of the Fiber Optic Coupled FTIR.

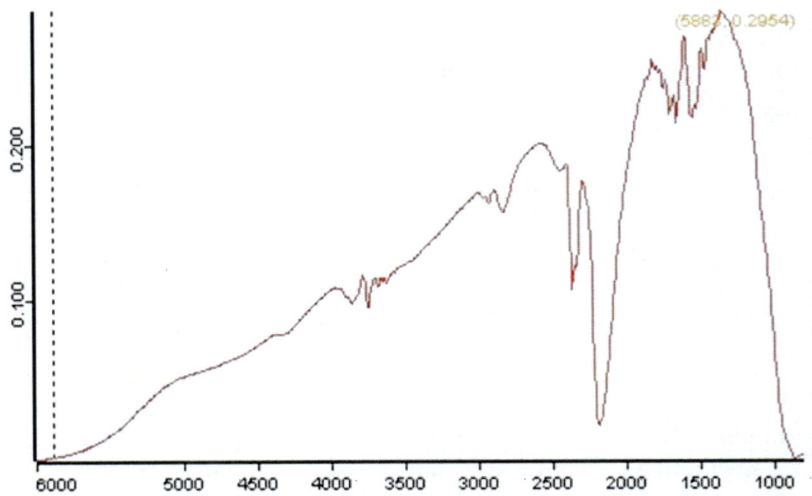

Figure 2. Transmission spectrum of chalcogenide glass used in the MIR optical fiber bundle used.

QUALITATIVE ANALYSIS OF SPECTRA

A typical IRRAS spectrum of the target analyte, designated as API1, is shown in Figure 3. The IRRAS spectrum of API1 is dominated by bands centered at about 1780, 1640 and 1525 cm^{-1}. The IR vibrational frequency assignment is outside the scope of this review, which deals with detection, quantification and discrimination analyses.

For the sake of comparison, the corresponding absorption spectrum of AP1 obtained by traditional transmission MIR using pressed KBr pellets and converted to absorbance mode is also shown in Figure 3. The general characteristics of the molecular infrared spectra are essentially retained when the molecule is absorbed on the metal surface. However, some subtle variations in the IRRAS spectrum occur, such as band shift to high frequencies and increased intensity or disappearance of some bands, which indicates that a preferred orientation was adopted by the thin layer of molecules absorbed on the surface.

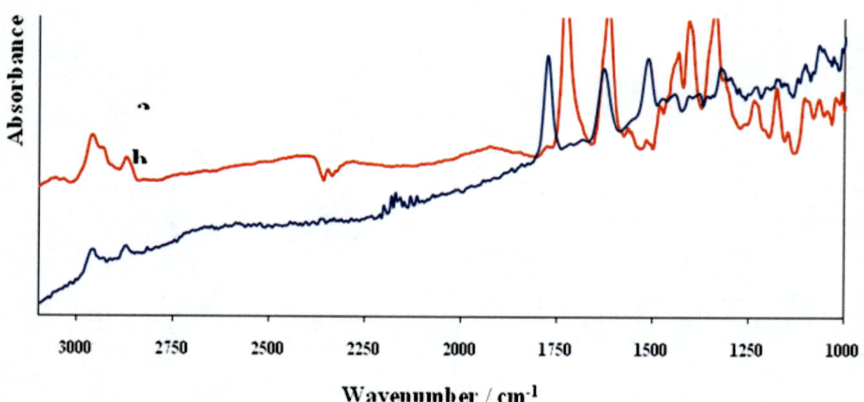

Figure 3. Spectra of API1 deposited on stainless steel. a) Bulk-KBr spectrum; b) IRRAS spectrum.

This phenomenon can be explained by applying the surface selection rule applicable to IRRAS spectroscopy and states that only those vibrations producing a dynamic dipole perpendicular to the surface will be observed [14].

DEVELOPMENT OF A CALIBRATION MODEL

Calibration models were built by using chemometrics-based, enhanced spectroscopy statistical tools, i.e., using the PLS1 algorithm of OPUS™ v. 4.2 (Bruker Optics, Billerica, MA) [15] as described in the above sections [16]-[17]. Nine hundred API1 samples were prepared by smearing sample transfer and analyzed spectroscopically as described above. During the calibration process, full cross-validation was applied using as many segments as samples included in the calibration set, which was done by using a "leave-one-out" method. In this approach, one spectrum is omitted from the training set and then tested against the model built with the remaining spectra. The process is then repeated with each of the spectra. The number of PLS components used to construct the models is determined from the lowest prediction error sum squares (PRESS) value. The overall predictive ability of each calibration model is assessed in terms of the root-mean-square error for the cross-validation (RMSECV) and prediction (RMSEP) sets. In all cases, none of the well-known spectral data pretreatments such as baseline correction, derivatives, spectral smoothing or multiplicative scatter correction were applied.

$$PRESS = \sum (\hat{C}_i - C_i)^2 \qquad \hat{C}_i: \text{Predicted concentration} \qquad (1)$$

C_i: True concentration

$$RMSEP = \sqrt{\frac{PRESS}{m_p}} \qquad m_p: \text{\# of samples not used in calibration} \qquad (2)$$

$$\text{RMSECV} = \sqrt{\frac{\text{PRESS}}{M}} \qquad \text{M: \# of samples used in cross-validation} \qquad (3)$$

The spectral range and the number of PLS factors are two of the most crucial parameters in the PLS1 regression process. All of the spectral data can be used to perform PLS studies; however, selecting a limited, highly significant spectral region is preferred in order to improve the prediction results. A cross validation in PLS1 within the training set was used to determine how many PLS components, or loadings, to include in the final model. The predicted values were then compared with the actual values for each of the calibration samples, and the PRESS values were calculated. The variance plot obtained using a spectroscopic window of 1810 to 1470 cm^{-1} for API1 is shown in Figure 4. The illustration shows that most of the variance is accounted for by the first two PCs, roughly 86.4%, and that 96.8% of the variance is contained in the first four components.

A detailed analysis of each of the loadings can facilitate a better understanding of the behavior of the factors in explaining the variance. Figure 5 shows the first five spectral components. The amplitudes of the loading spectra show their degree of covariance with the sample concentration. In this calibration, four loadings seemed to be significant because they presented common peaks with the grazing-angle FT-IR spectra of the API with positive or negative correlation. The fifth loading spectrum shows peaks uniformly distributed about zero; therefore, PC-5 and higher loadings that modeled only the noise may be considered. The four PLS factors that model absorbance and concentration of API were used later to reconstruct the spectrum.

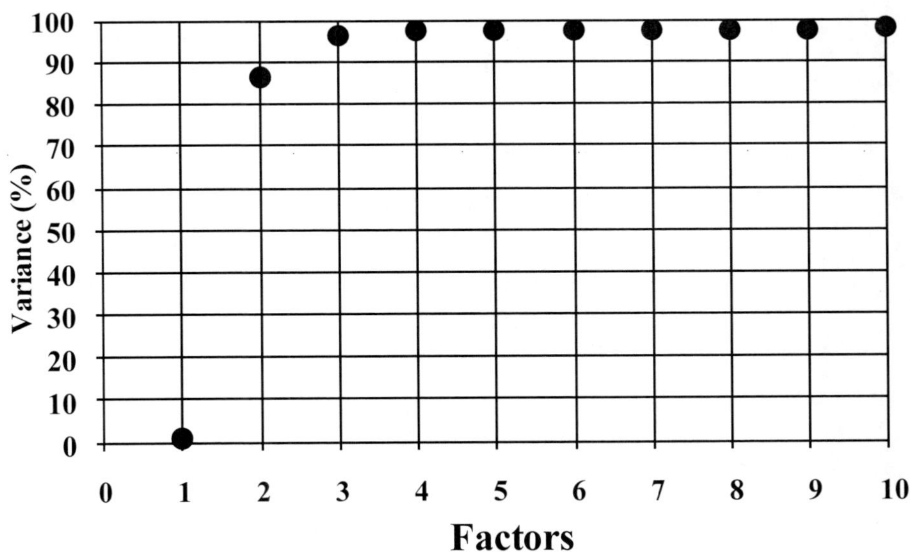

Figure 4. Plot of variance against number of PLS factors.

P. M. Fierro-Mercado, O. M. Primera-Pedrozo, A. Hornedo et al.

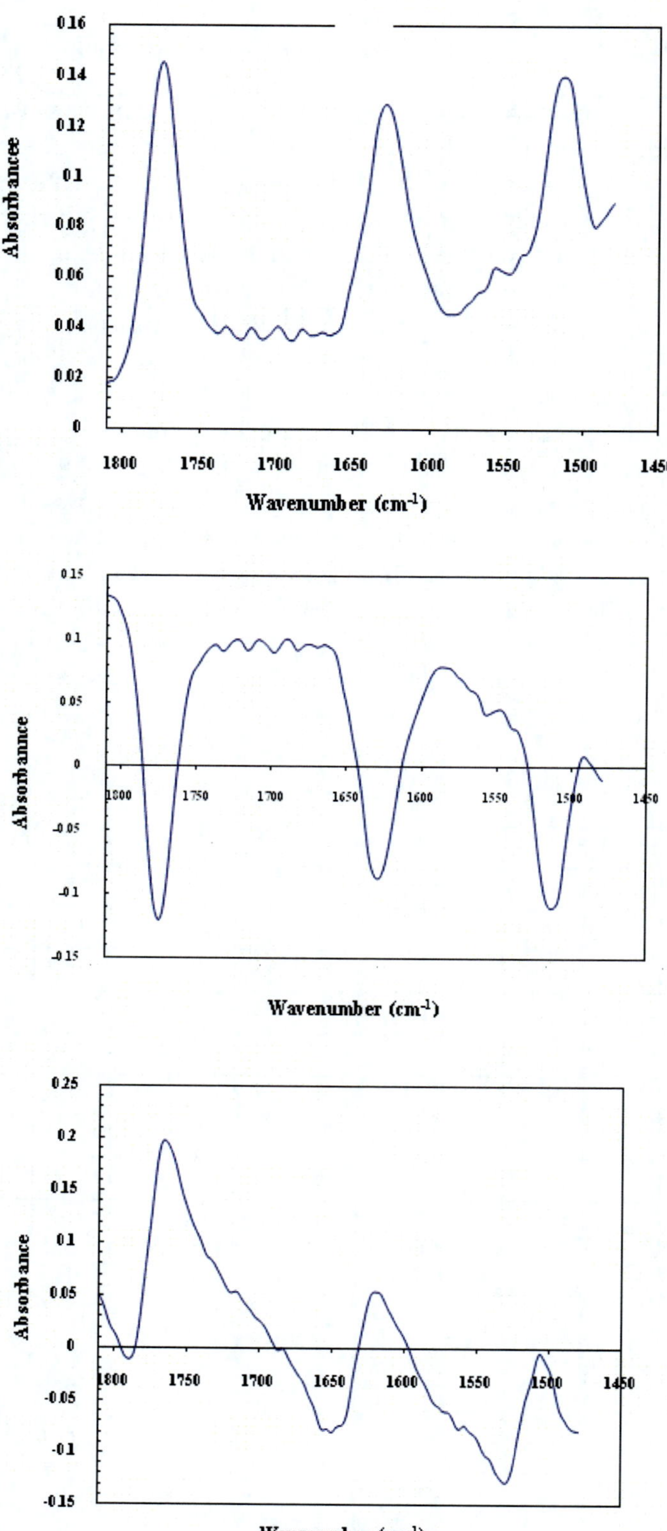

Figure 5 (Continued)

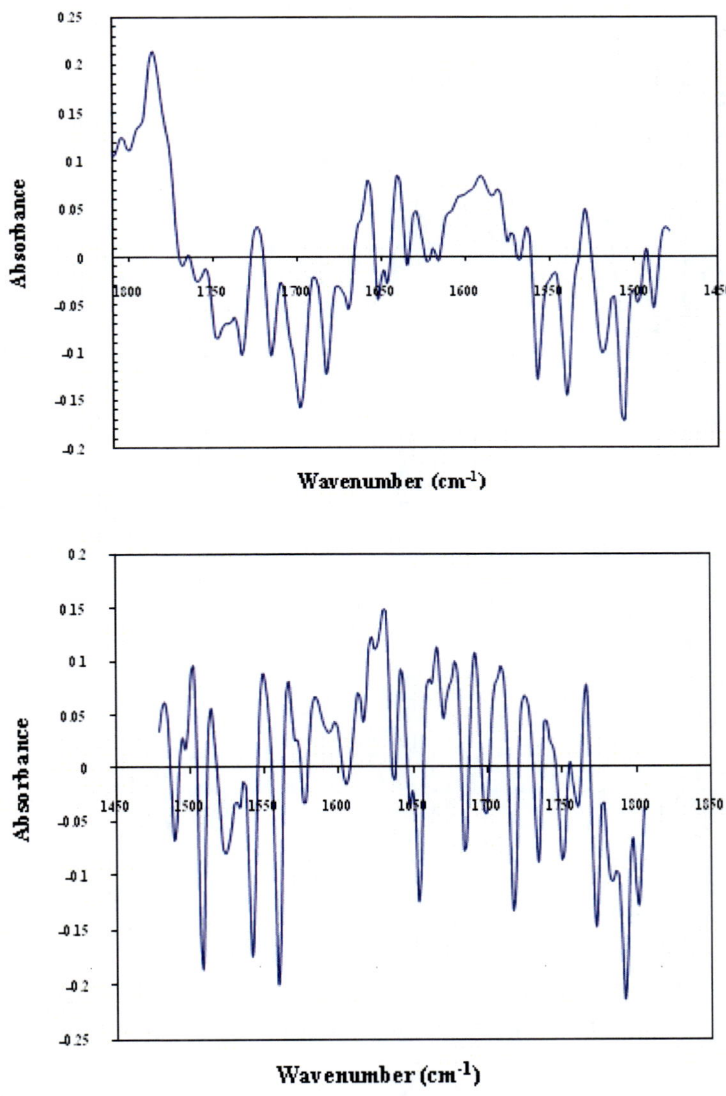

Figure 5. Spectral loadings for first five PLS factors: (a) first factor; (b) second factor; (c) third factor; (d) fourth factor; (e) fifth factor.

Figure 6 shows that these factors have an excellent modeling power of the IRRAS spectrum of the API studied.

Calibration models can be optimized by modifying the spectral region used for the analysis. After all of the data from the standards were included in a set of calibration data, the capacity of each generated model was evaluated. Based on the absorption features of the API spectra, two sub-ranges were investigated. The first region focuses on a subset of data including wavenumbers from 1810 to 1575 cm⁻¹, while the second one covers the subset from 1810 to 1740 cm⁻¹. Running PLS1 predictions on the two subsets and contrasting the results with the original PLS1 model provided the results summarized in Table 1.

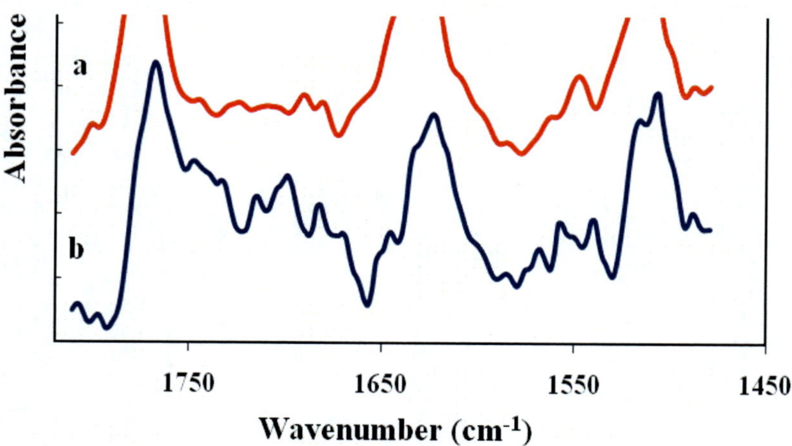

Figure 6. Grazing-angle FT-IR spectrum of API1. a) Original IRRAS spectrum; b) Modeled spectrum obtained with four PLS factors.

Table 1. Calibration and prediction results at various spectral regions

Region (cm^{-1})	Absorbance Spectra	
	RMSECV ($\mu g/cm^2$)	RMSEP ($\mu g/cm^2$)
1810.8 – 1479.1	0.422	0.495
1810.8 – 1575.6	0.545	0.530
1810.8 – 1739.4	0.502	0.499

The prediction accuracy is affected by restricting the wavenumber range to a narrower one. The data show that, despite a lower intensity than the 1750 cm^{-1} peak, the peaks at 1650 and 1500 cm^{-1} retain more information about the deposited concentration of the API even at lower concentration levels; however, this is not apparent by visual inspection of the spectra.

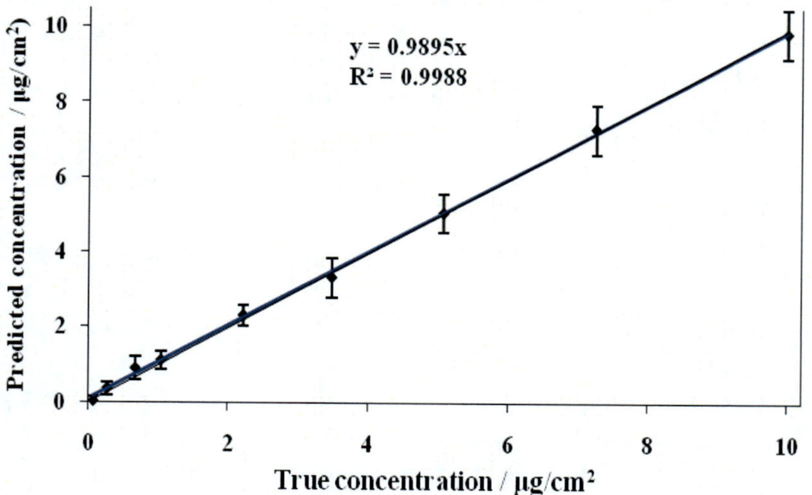

Figure 7. Calibration curve for API1 using 4 PLS factors; R^2 = 0.999.

With these criteria, calibration sets were prepared using known concentrations of API and plotting them against the concentrations calculated from the models, as illustrated in Figure 7. The statistical parameter regression coefficient squared (R^2) gives the fraction of variance present in the true component values, which are accounted for in the regression. The quality of the model can be judged from the R^2 value because the predicted concentration is in agreement with the true value when $R^2 \rightarrow 1$. In the case presented in Figure 7, the R^2 value was 0.9988, indicating a highly robust model. For models that best represent the real values, the slope tends to unity ($m \rightarrow 1$). In this model, the slope was 0.9895 and the intercept was zero.

An additional approach to test the robustness of the established model uses the external validation, which is developed measuring a subset of samples that are not part of the calibration set [19]. Table 2 shows the predicted concentrations of a subset of six samples with reference to the real values. The model used gave excellent predictions and could be used to detect concentrations below 40 ng/cm^2.

Table 2. Validation results for API1 using four-PLS-1 model

Sample	Amount Deposited (μg/cm^2)	Amount Detected (μg/cm^2)	Relative Error
1	0.36	0.43	0.19
2	0.62	0.73	0.18
3	0.91	0.98	0.08
4	1.46	1.45	1.0×10^{-3}
5	2.33	2.17	0.07
6	3.96	3.61	0.09

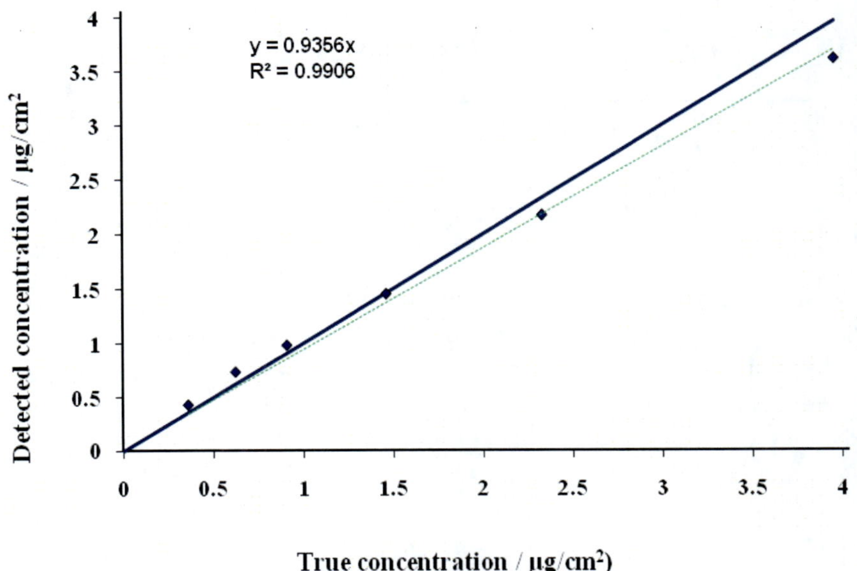

Figure 8. Validation results for API1 using four-PLS model.

The predicted concentrations show a slight linear deviation from the reference values when compared to a 45° line (m = 1), as shown in Figure 8. The resulting values of bias and slope, 0 and 0.9356, respectively, were used to correct the calibration model ($y_{corr} = a + by_{pred}$). The validation of the new corrected model was performed in a new set of samples with a resulting RMSEP of 0.136. Table 3 shows the results obtained after external validation using the PLS1 model compared with those obtained using the HPLC method as a reference method There are no appreciable differences between the two methods, which is an indication that this model is robust and can be used for both prediction and correlation.

The limit of detection is the smallest quantity of a substance that can be detected with reasonable certainty in the absence of the substance by a given analytical procedure [20]-[21]. An adequate method for determining the detection limit involves making an independent analysis of a suitable number of samples known to be near or prepared at the detection limit. For this API, samples ranging from 100 to 600 ng/cm^2 were deposited, and the concentrations were predicted using the 4-PLS model shown above. A detection limit of 280 ng/cm^2 was found to be acceptable, taking into account the lower standard error of prediction (SEP) and relative error obtained for this value (data not shown).

Table 3. Comparison of external validation of API1 using PLS and HPLC methods

Sample	Amount deposited ($\mu g/cm^2$)	Amount detected PLS-method ($\mu g/cm^2$)	Difference ($\mu g/cm^2$)	Amount detected HPLC-method ($\mu g/cm^2$)	Difference ($\mu g/cm^2$)
1	0.09	Not detected	------	0.09	0.00
2	0.29	0.23	-0.06	0.28	-0.01
3	0.40	0.26	-0.13	0.48	0.08
4	0.85	0.66	-0.19	0.87	0.02
5	1.15	1.02	-0.13	1.02	-0.13
6	2.00	2.08	0.08	1.93	-0.07
7	3.10	3.36	0.26	2.74	-0.36
8	4.40	3.91	-0.49	3.81	-0.59
9	6.10	5.99	-0.11	5.75	-0.35

Since the Grazing-angle FT-IR along with PLS has been shown to have sufficient modeling power to predict the concentration of API deposited on stainless steel surfaces, the effects of resolutions on these predictions were investigated. The vibrational bands related to the API spectrum at the region 1810 – 1000 cm^{-1} are visible at 1, 2, and 4 cm^{-1} resolution; however, some are not distinguishable at 8 and 16 cm^{-1} resolution. Table 4 shows the results for each resolution with PLS software. Using these conditions, we found that the standard error of prediction is lower for the spectra recorded at 4 cm^{-1} resolution.

This result shows that the increase in spectral information may induce over-fitting, which is prejudicial for the quantitative analysis. Based on the work capacity of the software, when more and more information is introduced into the model, there is a larger probability that the estimation process draws noise and other spurious phenomena from the calibration data into the resulting calibration model. Thus, the results of the calibration data appeared to correct; however, when used for prediction, they failed completely.

Table 4. Calibration and prediction results at various resolutions with PLS software

Resolution (cm^{-1})	Absorbance spectra	
	RMSECV	RMSEP
1	0.336	0.247
2	0.373	0.275
4	0.248	0.134
8	0.405	0.290
16	0.454	0.294

DISCRIMINANT ANALYSIS (DA)

Discriminant analysis (DA) is based in the classification of a set of observations into predetermined classes [22]. DA provides a rapid estimate of the identity and presence of residues left on surfaces at high or low levels according to the cleaning validation acceptance limits set by pharmaceutical manufacturing operations. Discriminant analysis is based on principal component analysis (PCA), in which the model makes calculations of its own vectors (eigenvectors, factors or loadings) that represent changes in the spectrum. In this way, it is possible to create a set of eigenvectors that characterize the changes in absorbance common to all spectra [23]. Once this mathematical treatment is performed, data are reduced to both a matrix of loadings and a matrix of scale constants, which are later used for reconstruction of the spectra. When this process is finalized, several functions are derived whose representation in the mathematical space shows groupings of objects with similar characteristics. Partial least squares regression (PLS) provides a decomposition routine in factors similar to PCA; however, PLS uses the information from the concentrations in the respond matrix.

Two discrimination models can be created; in the first, discriminant analysis is performed to classify the API loading concentration into one of three groups. The first group corresponds to no API concentration, the second group corresponds to concentrations lower than 800 ng/cm^2, and the third group corresponds to concentrations higher than 800 ng/cm^2. The threshold value corresponds to the maximum amount permitted for classifying the surface as clean and ready to use in batch changeover. Signals in the range of 1811-1479 cm^{-1} were used for the discrimination.

Decomposition in principal components (PC) of the total set of spectra was performed using Statgraphics for Windows™ 10.0, which found two functions for discrimination. Seven principal components contained 95% of the variance, and score analysis in this space discriminates the samples very well, although a few samples lie in intermediate positions, as shows in Figure 9. The best discriminant model was selected based on statistical significance and the percentage of cases classified correctly. The percentage of cases correctly classified was 98%, with a statistically significant p-value < 0.0001.

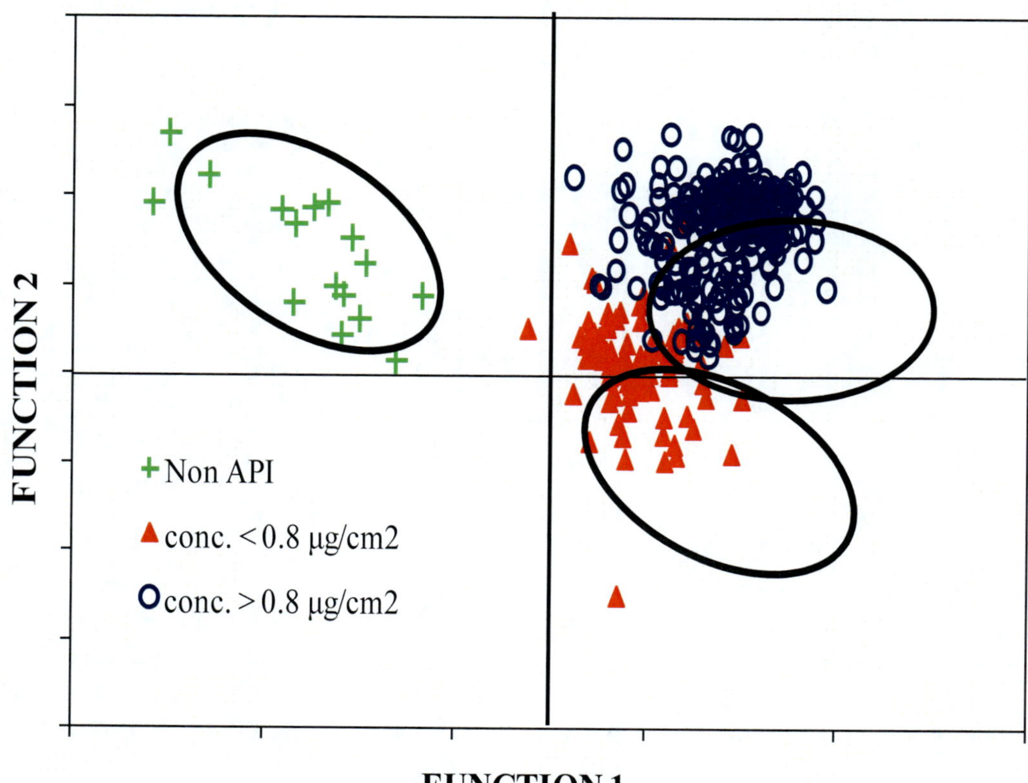

Figure 9. Discrimination model for API1 determination.

In the second discriminant model, a mixture of API1 and magnesium stearate (MgST), an excipient widely used in the pharmaceutical industry as a diluent and lubricating agent in the manufacture of medical tablets, capsules and powders, was prepared at concentrations between 0.1 and 5.0 $\mu g/cm^2$ and spectroscopically detected by the methodology as described above. As seen in Figure 10, the IRRAS spectrum of MgST shows bands that interfere with the detection of target surface contaminant since they occur in the region that contains the majority of signals from the API1. These results justify the use of a robust discriminant model to adequately overcome the overlapping of the IR signals that lead to nonlinear behavior reflected in variations in relative intensities and wavelength shifts as the surface loadings change with the complexity of the analysis matrix.

The data set consisted of a total of 1009 samples of the neat components: API1, MgST, binary mixtures of API1:MgST and no sample (blanks) in the range of 70 ng/cm^2 and 10 $\mu g/cm^2$. Figure 11 shows a cross-validation plot for API1 in the mixtures prepared with the value of loading concentrations for API and MgST, respectively, against the value predicted by the multivariate model in the global cross-validation. Similar plots were obtained for cross-validation of MgST in the presence of API1 (data not shown). Pre-processing was not used in any of the calibration runs.

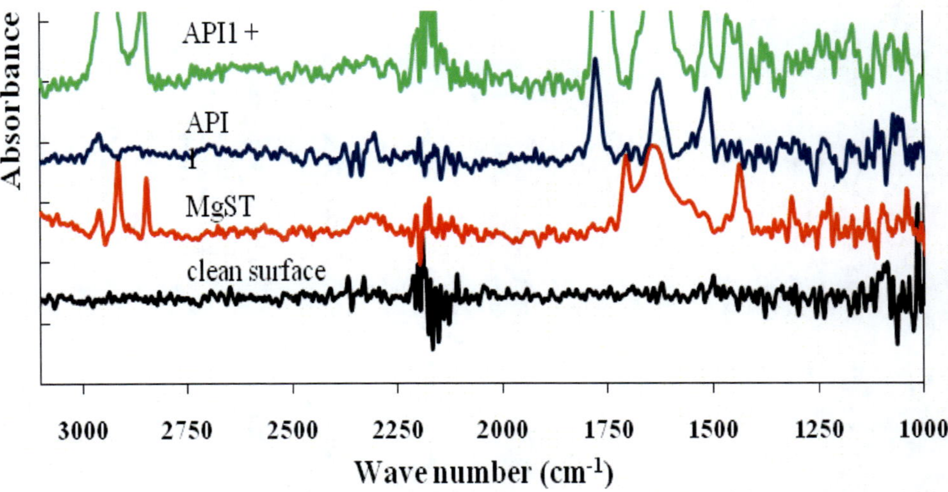

Figure 10. Grazing-angle FT-IR spectra of a mixture of BMS-API1 plus MgST, BMS-API1 and MgST alone, deposited on a stainless steel surface.

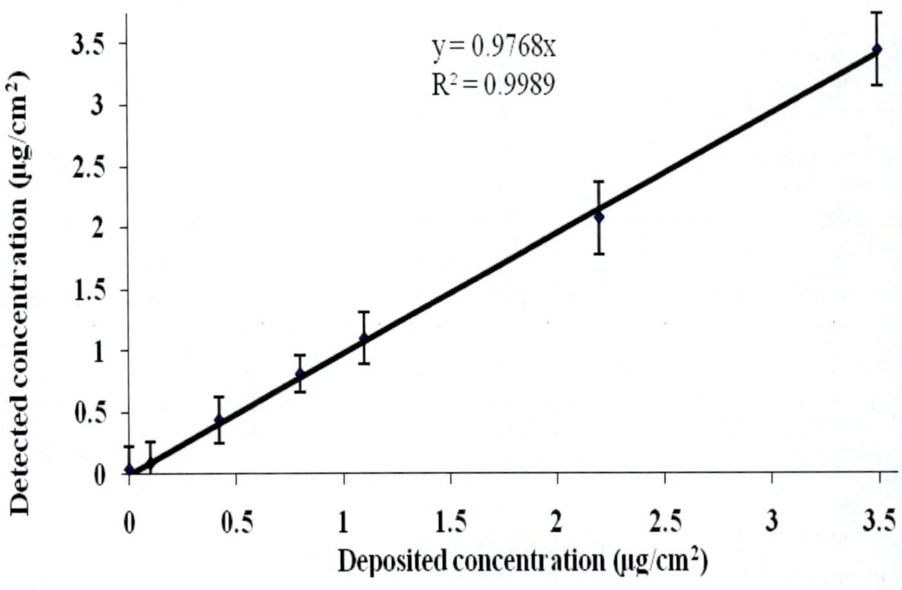

Figure 11. Cross-validation for a mixture of API1 and MgST using the PLS model with non-corrected data. PLS-factor number: 13.

In this study, five principal components contained 96% of the variance with a model built by concentrations higher than 1.0 µg/cm² (Figure 12). Concentrations below this value showed inadequate distributions, while 100% of the cases were classified correctly in the concentration range studied.

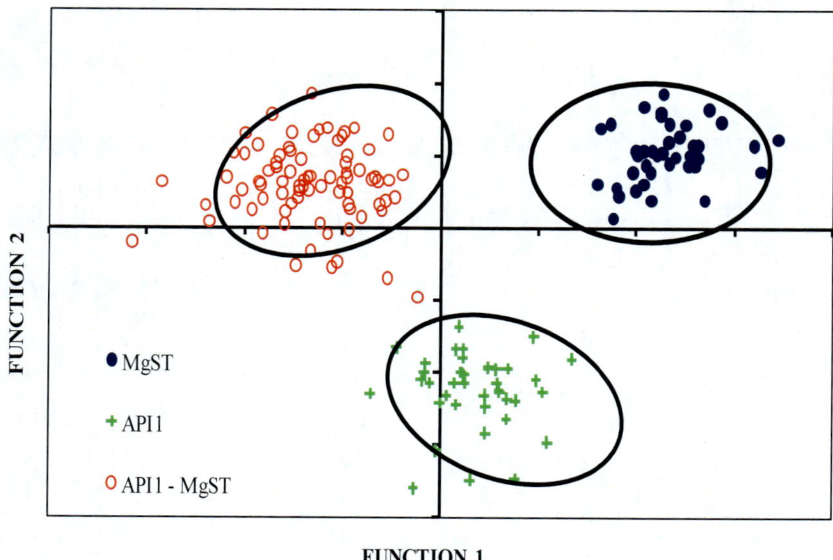

Figure 12. Discrimination model for API1, MgST and mixtures of API1 – MgST.

CONCLUSION

Grazing-angle probe Fourier Transform infrared spectroscopy with a partial least squares model can improve the current technique for the determination of organic matter on surfaces. This procedure appears to be a very promising method for rapid and low-cost determination of the presence of residues left on the surfaces of pharmaceutical reactors, contaminants on substrates, corrosion studies and quantification and discrimination studies of illicit substances, including explosives and drugs. It has been demonstrated that PLS and principal component analysis can adequately model data collected on chemicals adsorbed on surfaces for the purpose of predicting the surface concentration of the contaminants.

In the case studies presented, it was found that, in order to determine the information contained in the three maximum absorption peaks for an active pharmaceutical ingredient (API1) in the wavenumber region of 1800 to 1400 cm^{-1} of the spectra, the use of chemometrics routines for optimum modeling of the spectroscopic data was necessary. In addition, it was demonstrated for these data sets that high resolution FTIR was not crucial to ensure adequate prediction accuracy Classical least squares calibration techniques are not sufficient to discriminate the presence of more than one component in the sample if the spectroscopic signals overlap. This lack of sensitivity makes the univariate model less useful in many technical applications. On the other hand, multivariate calibration takes care of the different components in a complex sample. There is a significant advantage in using a factor-based method because it models data and organizes them based on the similarity of the information contained, which can also aid in a best interpretation of the system.

For the single-component contaminant on stainless steel surfaces, the calibration model obtained has a root-mean-square error of cross-validation (RMSECV) and prediction (RMSEP) of 0.422 and 0.495, respectively. Correction over a straight line was applied at an external test set, giving an RMSEP of 0.136. Thus, the PLS model used in this study had an excellent modeling capacity, taking into account the broad concentration range studied (70 ng/cm^2 to 10 µg/cm^2). An independent analysis of samples that ranged from 0.10 to 0.60 µg/cm^2 showed a detection limit of 280 ng/cm^2 with a relative error of 3.6%. In order to improve the prediction capacity in the multi-component system studied, mathematical treatments were applied; however, no differences were observed in comparison with data without pre-processing. For this reason, pre-processing was not used for the multicomponent analysis. The PLS model obtained shows an excellent linear correlation for both components, although no significant difference was found when compared to MgST at a low loading concentration. In this case, the model failed in predicting the concentrations of new samples.

ACKNOWLEDGMENTS

This work was done in collaboration with Bristol-Myers-Squibb (BMS), Humacao, PR. Thanks are due to BMS for supplying active pharmaceutical ingredients used in this investigation as well as magnesium stearate and stainless steel surfaces typically used in pharmaceutical batch reactors and other processing equipment.

Dan Klevisha (Bruker Optics, Billerica, MA), Peter J. Melling and Mary A. Thomson (Remspec Corporation, Sturbridge, MA) are gratefully acknowledged for their helpful suggestions and discussions on the use of the MIR fiber optic coupled-grazing angle probe.

REFERENCES

[1] Perston, B. B.; Hamilton, M. L.; Williamson, B. E.; Harland, P. W.; Thomson, M. A.; Melling, P. J. *Anal. Chem.* 2007, 79, 1231–1236.

[2] Hamilton, M. L.; Perston, B. B.; Harland, P. W.; Williamson, B. E.; Thomson, M. A.; Melling, P. J. *Org. Process Res. Dev.* 2005, 9, 337–343.

[3] Nozal, M. J.; Bernal, J. L.; Toribio, L.; Jiménez, J. J.; Martín, M. T. *J. Chromatogr. A.* 2000, 870, 69-71.

[4] Guide to Inspections Validation of Cleaning Processes. U.S. Food and Drug Administration (FDA): Rockville, MD, 1993.

[5] Primera-Pedrozo, O.; Soto-Feliciano, Y.; Pacheco-Londoño, L.; Hernandez-Rivera, S. *Sens. Imaging.* 2009, 10, 1–13.

[6] Primera-Pedrozo, O.; Rodríguez, N.; Pacheco-Londoño, L.; Hernández-Rivera, S. P. *Proceedings of SPIE.* 2007, 6542, 65423J.

[7] Melling, P. J.; Thomson, M. In: *The Handbook of vibrational spectroscopy.* Chalmers J. M.; Griffiths P. R. Eds.; Wiley: Chichester, 2002; Vol. 2, pp 1551-1559.

[8] Melling, P. J.; Shelley, P. U.S. Patent 6,310,348, 2001.

[9] Mehta, N. K.; Goenaga-Polo, J. E.; Hernández-Rivera, S. P.; Hernández, D.; Thomson, M. A.; Melling, P. J. *BioPharm.* 2002, 15, 36-42.

[10] Primera-Pedrozo, O.; Pacheco-Londono, L.; Ruiz, O.; Ramirez, M.; Soto-Feliciano, Y.; De la Torre-Quintana, L.; Hernandez-Rivera, S. *Proceedings of SPIE.* 2005, 5778, 543–552.

[11] Primera-Pedrozo, O. M.; Soto-Feliciano, Y.; Pacheco-Londoño, L. C.; Hernandez-Rivera, S. P. *Sensing and Imaging: An International Journal.* 2009, 10, 1–13.

[12] Primera-Pedrozo, O. M.; Soto-Feliciano, Y.; Pacheco-Londoño, L. C.; Hernandez-Rivera, S. P. *Sensing and Imaging: An International Journal.* 2008, 9, 27–40.

[13] Erk, N. *J. of Chromatography B.* 2003, 784, 195-201.

[14] Attard, G.; Barnes, C. Surfaces. Oxford University Press: Great Britain. 2006, pp 78-82.

[15] OPUS 4.2 Version. User Manual. Bruker Optics: Billerica, MA. 2003.

[16] Brereton, R.G. Applied Chemometrics for Scientists. John Wiley and Sons. N.Y. 2007.

[17] Beebe, K.; Pell, R.; Beth, M. Chemometrics: A Practical Guide. John Wiley and Sons. N.Y. 1998.

[18] Johnson, R. A.; Wichern, D. W. Applied Multivariate Statistical Analysis. Prentice-Hall: Englewood Cliffs, NJ. 1992.

[19] Prado-Fernandez, J. A.; Rodriguez-Vasquez, A.; Tojo, E.; Andrade, J. M. *Anal. Chim. Acta.* 2003, 480, 23-37.

[20] Thomsen, V.; Schatzlein, D.; Mercuro, D. *Spectroscopy* 2003, 18(12), 112-114.

[21] IUPAC. Commission on Spectrochemical and Other Optical Procedures for Analysis: Nomenclature, Symbols, Units and Their Usage in Spectrochemical Analysis-II. Data Interpretation. *Pure Appl. Chem* 1976. 45, 99.

[22] Huberty, C. J. Applied Discriminant Analysis. Wiley–Interscience: NJ. 1994.

[23] Lavine, B. *Anal. Chem.* 2000, 72, 91R-97R.

In: Fourier Transform Infrared Spectroscopy
Editor: Oliver J. Rees, pp. 195-204

ISBN: 978-1-61668-835-6
© 2010 Nova Science Publishers, Inc.

Chapter 9

STRUCTURAL MODIFICATION OF SINGLE WALL CARBON NANOTUBES BY GAMMA IRRADIATION

*Biljana Todorović Marković * and Zoran Marković*
Vinča Institute of Nuclear Sciences, 11001 Belgrade, Serbia

ABSTRACT

One of the most versatile analytical chemical techniques is Fourier transform infrared (FTIR) spectroscopy. This technique is increasingly used for quantitative and qualitative analyses in many diverse applications. Among these are the analysis of pharmaceuticals, biomembranes, biopolymers and microbiological applications.

In this chapter, we have described the study of structural modification of single wall carbon nanotubes (SWCNTs) caused by γ irradiation. SWCNTs were irradiated in three different media: water, air and 30 % ammonia solution. Different techniques were used for investigation of modification of SWCNTs: Raman spectroscopy, FTIR spectroscopy and atomic force microscopy.

FTIR spectroscopy were used to examine covalent modifications of nanotubes' sidewalls. This technique has verified the functionalization of SWCNTs by detecting the presence of amino and hydroxyl groups at the sidewalls of SWCNTs.

1. INTRODUCTION

Carbon nanotubes (CNTs) are hollow cylindrical molecules consisting of single or many sheets of graphite wrapped into cylinders with ranging from a few nm to hundreds of nm [1]. Depending on the diameter and chirality (the chiral angle between hexagons and the tube axis), single wall carbon nanotubes can be metals, semiconductors and semimetals. They have attracted worlwide attention due to their unique mechanical, thermal and electronic properties [2]. They could find potential application in many areas like electronics, biotechnology, medicine, pharmacology [3-9]. However, their properties strongly depend on the extent of

---

* Corresponding author.e-mail:biljatod@vinca.rs

carbon nanotube dispersion and strenth of the interfacial adhesion, because CNTs form strongly bound aggregates due to their very large surface areas and strong van der Waals interactions. The most common way of processing of this material into a more useful state, either for characterization or to facilitate the development of applications, is to disperse it in a liquid. In recent years this has been achieved mainly by using third-phase dispersants like salts [10], surfactants [11,12], polymers [13-15] or biomolecules [16,17]. Surfactant-assisted dispersion is one of the possible methods used to obtain a stable dispersion [18-20]. It is observed that in chemical modification of nanotubes, functional groups are attached to defects of the SWCNT walls since the undamaged walls are chemically inert. Noncovalent functionalization of nanotubes allows us to tailor their properties, preserving almost all the original characteristics of CNTs, including the sp^2 nanotube structure and electronic properties. One of the most common ways to covalently functionalize carbon nanotubes is oxidation [21]. Another widely used type of covalent reaction to functionalize carbon nanotubes is the cyloaddition reaction which occurs on the aromatic sidewalls instead of nanotubes ends and defects as in the oxidation case [22].

Carbon nanotubes can be chemically modified by means of energetic particles as well. Until now there are only a few reports about irradiation effects on carbon nanotubes. Recent experiments have demonstrated that irradiation of carbon nanotubes with energetic particles-electrons and ions- can be used to create molecular junctions between carbon nanotubes and composite materials [23]. Khare et al. reported that C-H bonds formed on surface of SWCNTs exposed to 1 MeV proton irradiation [24]. The results indicate that proton irradiation to fluences as high as 5.6×10^{15} cm^{-2} has little effect on the interband transitions in carbon nanotubes [25]. As for electron beam irradiation of single wall carbon nanotubes, it was found that SWCNTs with small diameters of 1 nm are damaged by electron beam while SWCNTs with diameters of 1.3 nm and larger are more stable agains degradation and stability increases with diameter [26].

γ-ray irradiation is an effective and facile technique for surface modification of either organic or inorganic materials. Gamma rays are electromagnetic radiation of high frequency. They are produced by sub-atomic particle interactions such as electron-positron annihilation, neutral pion decay, radioactive decay, fusion, fission or inverse Compton scattering in astrophysical processes. Gamma rays typically have frequencies above 10^{19} Hz and therefore energies above 100 keV and wavelength less than 10 picometers, often smaller than an atom. Guo et al. concluded that the solubility of γ-irradiated multiwall nanotubes in organic solvents was considerably enhanced compared to unirradiated nanotubes [27]. Skakalova et al. found that defect formation after γ- irradiation could lead to cross-links between the SWCNTs in bundles and electrical conductivity of irradiated nanotubes was enhanced by factor of 4.5 [28]. The generation of defects on SWCNTs is in fact a physical functionalization process that may result in a better anchoring of SWCNTs within the polymeric matrix in polymer-carbon nanotube composites. Hulman et al. established that the G-band intensity increase is ascribed to a softening of the q=0 selection rules, rather than to the initializing of q>0 phonon scattering of the double-resonance model [29].

In this study we have investigated the effect of gamma irradiation on nanotube structure. This paper focuses mainly on the synthesis and characterization of stable colloidal suspension of SWCNTs. The critical elements for development of transparent conductive SWCNT electronics are highly stable colloids with minimum amount of stabilizing polymer and highly hydrophillic substrate [30]. In this work we chose melamine sulfonate superplasticiser (MSS) to disperse SWCNTs. Our previous results concerning pristine nanotube functionalization by MSS had shown that this organic polymer was successfully dispersed nanotubes. Melamine sulfonate superplasticiser is an organic polymer. The long molecules wrap themselves around the nanotubes, giving them a highly negative charge so that they repel each other. Anion groups binding to polymer backbone are fitted in the sidewall defects of SWCNTs. This polymer disperses the nanotubes through a mechanism of electrostatic repulsion. Gamma irradiation was performed in the three different media: water, air and aqueous solution of ammonia. FTIR spectra have shown the presence of amino groups after irradiation in all irradiated specimens. The amount of used polymer for functionalization was one order of magnitude less than that necessary for debundling pristine tubes [31]. It was established that Raman spectra of irradiated and functionalized nanotubes was differed from those of pristine and functionalized ones.

2. EXPERIMENTAL PROCEDURE

Pristine SWCNTs (diameter 5 nm, length 4 μm, purity > 95 %, purchased from Bucky USA) were irradiated with ^{60}Co in three different media: water, air and 30% aqua solution of ammonia (further-aqua ammonia). Irradiation was performed on a ^{60}Co irradiator (22.57 GBq in total activity at the Center of Irradiation, Vinča Institute of Nuclear Sciences) with photon energy of 1.3 MeV. The irradiation dose was 50 kGy.

Irradiated and pristine single wall carbon nanotubes (50 mg, 95% purity from BuckyUSA) were added to 200 ml of stock solutions, respectively. Stock solutions have been prepared by adding 2 mg melamine sulfonate superplasticiser to 200 ml of distilled/deionized water. The SWCNT solutions were sonicated (ultrasonic bath with power 750 W) for 3 hours. After sonication, samples were centrifuged at 4000 rpm (rcf=2575 xg) for 1 h. For each sample the well-dispersed supernatant was taken for characterization leaving behind the precipitate.

Pristine and γ-irradiated SWCNTs were dispersed in THF as well. Further in the text γ-irradiated nanotubes were designated as γNSWNT (SWCNTSs irradiated in aqua solution of ammonia), γHSWCNT (SWCNTs irradiated in water) and γOSWNT (SWCNTs irradiated in air), respectively.

The concentrations of all nanotube dispersions were determined by gravimetric method: 10 ml of nanotube dispersion was dried at 60 °C in air and the mass was measured. The concentration of pristine SWCNT/MSS suspension was 40 mg/L while the concentrations of γNSWCNT/MSS, γHSWCNT/MSS and γOSWCNT/MSS suspensions were 250 mg/L, 210 mg/L and 225 mg/L, respectively.

Raman spectra of pristine, irradiated and modified single wall carbon nanotubes were obtained by Micro Raman Chromex 2000 using 532 nm of a frequency doubled Nd:YaG laser with power of 2 mW. The spectral resolutions were 4 cm^{-1}. Raman spectra were recorded at room temperature. Samples of dried modified nanotubes were pressed in the shallow hole of indium substrate.

For the Fourier transform infrared (FTIR) spectroscopy analysis, nanotube suspensions were dried at temperature of 60°C. Dried nanotubes were mixed with KBr powder and pellets were formed. FTIR spectra were measured at room temperature in the spectral range from 400 to 4000 cm^{-1}, on a Nicollet 380 FT-IR, Thermo Electron Corporation spectrometer.

AFM measurements were performed using Quesant microscope operating in tapping mode in air at room temperature. In tapping mode the cantilever oscillates close to resonance and the tip only slightly touches the surface [32]. Mica was used as a substrate.

Diluted dispersions of nanotubes were deposited on mica substrate and imaged after drying. Standard silicon tips (purchased from Nano and more) with force constant 40 N/m were used. The accuracy of the AFM mean diameter determination was improved by deconvolution. To reduce the aggregation on the substrate dispersions were diluted 40 times.

3. RESULTS AND DISCUSSION

3.1. Raman Spectroscopy

Raman spectroscopy is powerful and non-destructive tool to characterize sp^2-bonded carbons, e.g. graphite, fullerenes and carbon nanotubes [33]. The differences in the Raman spectra of sp^2 carbons stem mainly from first-order double resonance scattering that are sensitive to the impact of symmetry on the electron and phonon states in the respective carbons which can be the manifestation of lateral quantum confinement (nanoribbons) or cylindrical boundary conditions (nanotubes), or the addition of pentagonal rings (fullerenes) [34].

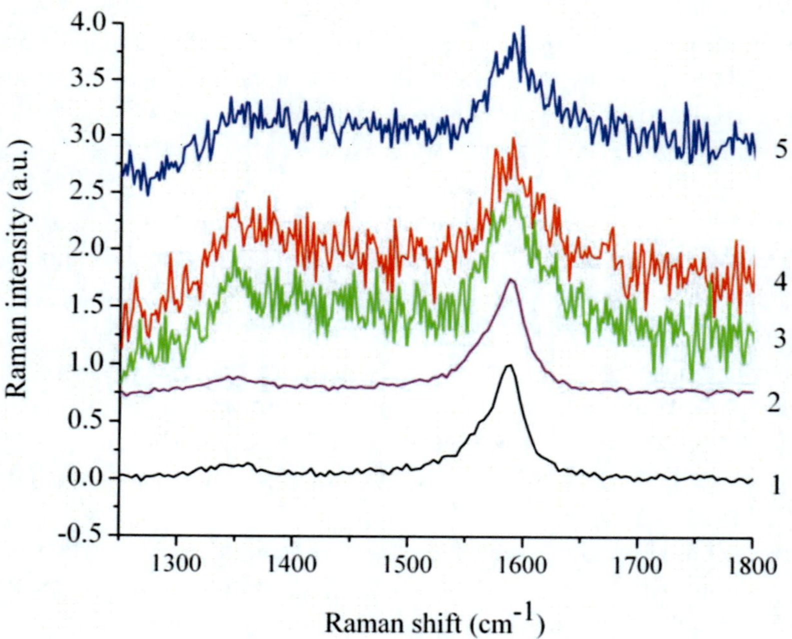

Figure 1. Raman spectra of pristine (curve 1), pristine, functionalized (curve 2), irradiated and functionalized SWCNTs in air (curve 3), aqua ammonia (curve 4) and water (curve 5).

In Figure 1, Raman spectra of pristine, irradiated and functionalized specimens are presented. The laser energy was 2.33 eV. As could be seen from Figure 1 two characteristic Raman bands are observed: a relatively broad band near 1300 cm^{-1} and a stronger band with structure in the 1550-1600 cm^{-1} [35].

The band with maximum near 1300 cm^{-1} is a common in disoredered sp^2 carbon matrix and has been called D-band. It is activated by disorder in sp^2 carbon network.

The band at 1590 cm^{-1} is close to that observed in well ordered graphite and is called G-band. As for D-band, it could be noticed from I_D/I_G ratio (Table 1) that there was significant increase of intensity of D-band after γ irradiation.

Table 1. I_D/I_G ratios and the positions of maximums of D and G bands respectively

	I_D/I_G	D band (cm^{-1})	G band (cm^{-1})
Pristine SWCNT/MSS	0.162	1361.2	1589
SWCNT/MSS	0.238	1344.1	1587.9
γNSWCNT/MSS	0.787	1378.1	1589.5
γHSWCNT/MSS	0.783	1344.6	1585.7
γOSWCNT/MSS	0.762	1348.6	1597.2

As could be observed from Table 1, I_D/I_G ratio increased almost three times. It is known that γ irradiation produces defects in the graphite lattices. Telling reported that vacancies produced by γ irradiation are stabilized by creating pentagon-heptanon defects and pushing one carbon atom out of the graphene plane [36]. Subsequently, a cross-link between neighboring graphene layers is formed if two displaced atoms meet each other. In our study gamma irradiation process causes significant increase of the number of defects in nanotube walls. In that way, better functionalization of used SWCNTs by MSS was enabled.

3.2. FTIR Spectroscopy

FTIR spectroscopy is primarily a qualitative tool used to identify functional groups and the nature of their attachment to CNT sidewalls. Different functional groups absorb characteristic frequences of infrared radiation, giving rise to a fingerprint identification of bonds. In Figure 2, FTIR spectra of pristine single wall carbon nanotubes, melamine sulfonate superplasticiser, gamma irradiated nanotubes in air, aqua and aqua ammonia and functionalized by MSS are presented.

As could be seen from diagram, the peaks at 1035, 1169 and 1385cm^{-1} (presented in curves 2,3,5,6) are assigned to ionic sulfonate SO_3^- group [37]. The presence of these peaks indicates the functionalization of pristine and irradiated SWCNTs by MSS. The peaks at 2850 and 2920 cm^{-1} (presented in all specimens) indicate the presence of C-H bonds while broad band at about 3400 cm^{-1} belongs to OH groups [37]. The presence of broad band at 650-680 cm^{-1} only in irradiated specimens obviously shows that amino groups have been introduced onto the surface of irradiated nanotubes by gamma irradiation. The peaks which were detected in irradiated and functionalized specimens could not be observed in pristine nanotubes (as-produced by Bucky USA). Based on these data we could conclude that gamma irradiation introduces significant change in the structure of nanotubes. Further, the presence of amino and hydroxyl groups in the FTIR spectra of irradiated specimens enables better functionalization of nanotubes and their debundling as well. The amount of used polymer (MSS) was about 10 times lower than for non irradiated and functionalized nanotubes [31].

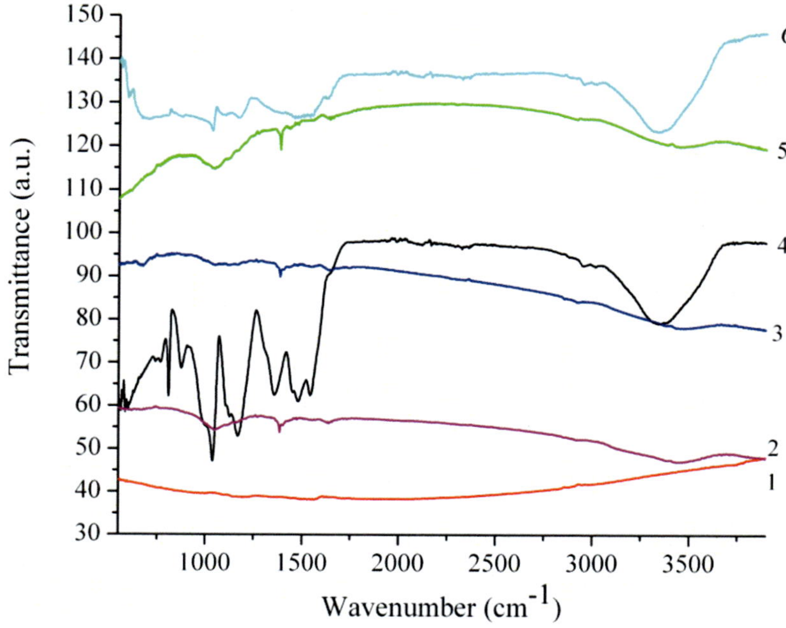

Figure 2. FTIR spectra of pristine (red curve 1), functionalized (cyan curve 6) and γ-irradiated and functionalized by MSS in air (magenta curve 2), ammonia (blue curve 3) and water (green curve 4). FTIR spectrum of MSS is also presented (black curve 5).

3.3. Atomic Force Microscopy

The distribution of lenghts and diameters of pristine, irradiated and functionalized SWCNTs were characterized by AFM operating in tapping mode. Top view AFM image of irradiated nanotube bundle (irradiation dose – 50 kGy) dispersed in THF is presented in Figure 3a. Average size of nanotube bundle is about 20 nm. This value is much smaller than that obtained for non-irradiated tubes dispersed in THF [31] and could indicate better solubility of irradiated tubes in tetrohydrofuran. Top view AFM images of pristine and irradiated and functionalized nanotube bundles are presented in Figs. 3 (b,c,d,e). Non-irradiated and modified carbon nanotubes by MSS are bent likely as a result of ultrasonic treatment-Figure 3b. The presence of micelles could be clearly observed in Figure 3b. These micelles are not presented in irradiated and functionalized specimens. The average size of nanotube bundle irradiated in air was 30 nm-Figure 3c while those irradiated in aqua ammonia were 15 nm-Figure 3d.

The average size of nanotube bundle irradiated in water was 40 nm. The irradiation medium affects the nanotube debundling as well. SWCNTs irradiated in aqua ammonia have shown the best nanotube debundling. We suppose that introduction the OH and amino groups on nanotube surface attributed better nanotube debundling comparing to other used media. FTIR and Raman results confirm this presumption.

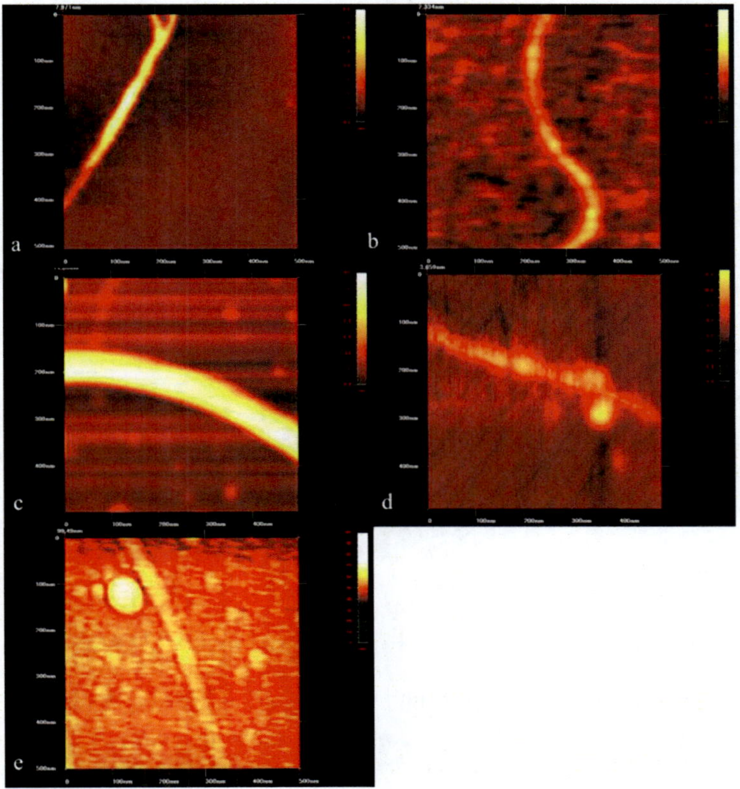

Figure 3. Top view AFM images of pristine (a), functionalized by MSS (b), γ-irradiated and functionalized nanotubes in air (c) aqua ammonia (d) and water (e). The irradiation dose was 50 kGy.

According to AFM images we are able to observe single wall carbon nanotubes within their bundles and determine their average size. The presence of micellas in nanotube dispersion was very small because the amount of used polymer was very small (polymer:nanotube=1:25). AFM results are in a good correspondence to measured concentration values of pristine and γ-irradiated specimens.

3.4. Stabilization Mechanism of Nanotube Dispersions

Melamine sulfonate superplasticiser is an organic polymer and could be dispersed in aqueous media. This polymer has sodium sulfonate group and dispersion of pristine nanotubes is achieved by the interaction between inner hydrophobic site of the polymer and hydrophobic surface of SWCNTs together with a suitable orientation of outer sodium sulfonate group to water. Possible mechanism of stabilization of gamma irradiated nanotubes and functionalized by MSS is presented in Figure 4. As could be seen from the Figure 4, NH_2 and OH groups are covalently bonded to sidewall of SWCNTs. SO_3^- anions binding to polymer backbone are fitted in the sidewall defects of SWCNTs while melamine polymer backbone itself is wrapping around nanotubes and π-π interactions are established between aromatic rings and graphitic tube surface [38,39]. The π-π interaction between the polymers and the SWCNTs surfaces contributes to the solubilization of the SWCNTs [40].

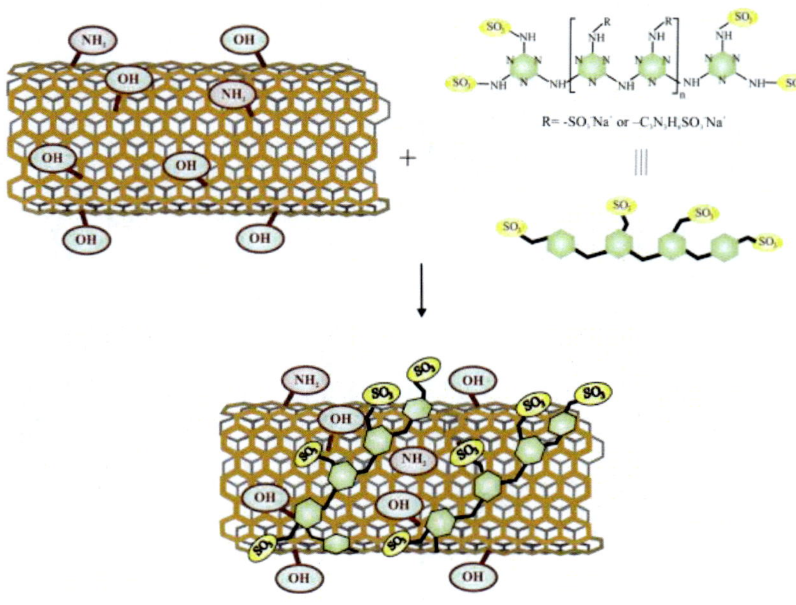

Figure 4. Reaction scheme for the modification of γ-irradiated SWCNTs by MSS.

Main advantage of irradiated tubes that promotes their dispersion is the presence of NH_2 and OH groups which were covalently bonded to their sidewall. But the number of these groups is not enough to enable nanotube separation without the help of polymer such as MSS. Polymer backbone is wrapped around the nanotube surface and between amino and OH groups. According to Raman results, the number of defects introduced by gamma irradiation is highly increased that enables better dispersion of nanotubes.

In the case of noncovalent functionalization, carbon nanotubes solubilized by those amphiphiles with relatively high critical micelle concentration (CMC) are typically not stable without an excess of surfactant molecules in the solution. In our study, the amount of used polymer (2 mg) was not enough to hold stable pristine nanotube/MSS dispersion for a long time. On the contrary, γ-irradiated nanotubes were dispersed by MSS and this colloid was stable for long period. Furthermore, the amount of used organic polymer for irradiated nanotube debundling is one order of magnitude lower than that used for debundling of pristine nanotubes.

In order to verify the stabilization of pristine and γ-irradiated nanotube dispersions by electrostatic repulsive forces we caused the precipitation of nanotube dispersion by adding 1 wt.% NaCl solution. The precipitation was occurred during several hours.

CONCLUSION

In this work we have investigated the effect of gamma irradiation on SWCNT structure. Several different techniques have been used for characterization of pristine and irradiated nanotubes:Raman spectroscopy, FTIR spectroscopy and atomic force microscopy. It was found that gamma irradiation enhances the number of defects incorporated in the sidewalls of SWCNTs. NH_2 and OH groups are covalently bonded to nanotube's sidewalls. Functionalization of nanotubes by MSS after their gamma irradiation in three different media attributes to their better debundling. The amount of used polymer for nanotube functionalization is one order of magnitude lower than that used for functionalization of pristine SWCNTs.

ACKNOWLEDGMENTS

This research was supported by the Ministry of Science of Republic of Serbia (project no. 141015).

REFERENCES

[1]   Dresselhaus, M.S.; Dresselhaus, G.; Eklund, P. C. Science of Fullerenes and Carbon Nanotubes; Academic Press: San Diego, CA, 1996.

[2]   Saito, R,; Dresselhaus, G.; Dresselhaus, M. S. Physical Properties of Nanotubes; Imperial College Press: London, UK, 1998.

[3]   Li, F.; Cheng, B. S.;Su, G.; Dresselhaus, M. S. *Appl. Phys. Lett.* 2000, 77, 3161-3163.

[4]   Smith, B. W.; Benes, Z.; Luzzi, D. E.; Fisher, J. E.; Watters, D. A.; Casavant, M. J.; Schmidt, J.; Smalley, R. E. *Appl. Phys. Lett.* 2000, 77, 663-665.

[5]   Klumpp, C.; Kostarelos, K.; Prato, M.; Bianco, A. *Biochim. Biophys. Acta* 2006, 1758, 404-417.

[6]   Lacereta, L.; Bianco, A.; Prato, M.; Kostarelos, K. *Adv. Drug Delivery Rev.* 2006, 58, 1460-1470.

[7]   Foldvari, M.; Bagonluri, M. *Nanomed. Nanotechnol. Biol. Med.* 2008, 4, 173-182.

[8]   Singha, N.; Yeow, J. *IEEE Trans. Nanobiosci.* 2005, 4, 180-195.

[9]   Raffa, V.; Ciofani, G.; Nitadas, S.; Karachalios, T.; D'Alessandro, D.; masini, M.; Cuschiori, A. Carbon, 2008, 46, 1600-1610.

[10] Penicaud, A.; Poulin, P.; Derre, A.; Anglaret, E.; Petit, P. *J. Am. Chem.* Soc. 2005, 127, 8-9.

[11] Moore, V. C.; Strano, M. S.; Haroz, E. H.; Hauge, R. H.; Smalley, R. E.; Schmidt J.; Talmon, Y. *Nano Lett.* 2003, 3, 1379-1382.

[12] Islam, M. F.; Rojas, E.; Bergey, D.M.; Johnson, A. T.; Yodh, A. G. *Nano Lett.* 2003, 3 269-273.

[13] Coleman, J. N.; Dalton, A. B.; Curran, S.; Rubio, A.; Davey, A. P.; Drury, A.; McCarthy, B.; Lahr, B.; Ajayan, P. M.; Roth, S.; barklie, R. C.; Blau, W. J. *J. Adv. Mater.* 2000, 12, 213-216.

[14] O'Connell, M. J.; Boul, P.; Ericson, L. M.; Huffman, C.; Wang, Y. H.; Haroz, E.; Kuper, C.; Tour, J., Ausman, K. D.; Smalley, R. E. *Chem. Phys. Lett.* 2001, 342, 265-271.

[15] Liu, Y. Q.; Gao, L.; Zheng, S.; Wang, Y.; Sun, J.; Kajiura, H.; Li, Y.; Noda, K. *Nanotechnology*, 2007, 18, 365702.

[16] Zheng, M.; Jagota, A.; Semke, E. D.; Diner, B. A.; Mclean, R. S.; Lustig, S. R.; Richardson, R. E.; Tassi, N. G. *Nat. Mater.* 2003, 2, 338-342.

[17] Zheng, M.; Jagota, A.; Strano, M. S.; Santos, A. P.; Barone, P.; Chou, S. G.; Diner, B. A.; Dresselhaus, M. S.; Mclean, R. S.; Onoa, G. B.; Samonidze, G. G.; Semke, E. D.; Usrey, M.; Walls, D. *J. Science*, 2003, 302, 1545-1548.

[18] Ham, H. T.; Choi, Y. S.; Chung, I. J. *J. Colloid Interface Sci.* 2005, 286, 216-223.

[19] Tasis, D.; Tagmatarchis, N.; Georgakilas, V.; Prato, M. *Chem. Eur. J.*, 2003, 9, 4000-4008.

[20] Star, A.; Steuerman, D. W.; Heath, J. R.; Stoddart, J. F. *Angew. Chem. Int. Ed.* 2002, 41, 2508-2512.

[21] Niyogi, S.; Hamon, M. A.; Hu, H.; Zhao, B.; Bhowmik, P.; Sen, R.; Itkis, M. E. Haddon, R. C. *Acc. Chem. Res.*, 2002, 35, 1105-1113.

[22] Georgakilas, V.; Kordatos, K.; Prato, M.; Guldi, D. M.; Holzinger, M.; Hirsch, A. *J. Am. Chem. Soc.*, 2002, 124, 760-761.

[23] Krasheninnikov, A. V.; Nordlund, K.; Keinonen, J.; Banhart, F. *Nucl. Instrum. Methods* B, 2003, 202, 224-229.

[24] Khare, B.; Meyyappan, M.; Moore, M. H.; Wilhite, P.; Imanaka, H.; Chen, B. *Nano Lett.*, 2003, 3, 643-646.

[25] Neupane, P. P.; Manasreh, M. O.; Weaver, B. D.; Rafaelle, R. P., Landi, B. *J. Appl. Phys. Lett.*, 2005, 86, 1-3.

[26] Warner, J. H.; Schaffel, F.; Guofang, Z.; Rummeli, M. H.; Buchner, B.; Robertson, J.; Briggs, G. A. D. *ASC Nano*, 2009, 3, 1557-1563.

[27] Guo, J.; Li, Y.; Wu, S.; Li, W. Nanotechnology, 2005, 16, 2385-2388.

[28] Skakalova, V.; Dettlaff-Weglikowska, U.; Roth, S. *Diamond Relat.Mater.*, 2004, 13, 296-298.

[29] Hulman, M.; Skakalova, V.; Roth, S.; Kuzmany, H. *J. Appl. Phys.*, 2005, 98, 1-5.

[30] Gu, H.; Swager, M. *Adv. Mater.*, 2008, 20, 4433-4437.

[31] Marković, Z.; Jovanović, S.; Kleut, D.; Romčević, N.; Jokanović, V.; Trajković, V.; Todorović Marković, B. *Appl. Surf. Sci.*, 2009, 255, 6359-6366.

[32] Todorović Marković, B.; Jovanović, S.; Jokanović, V.; Nedić, Z.; Dramićanin, M.; Marković, Z. *Appl. Surf. Sci.*, 2008, 255, 3283-3288.

[33] Graupner, R. *J. Raman Spectrosc.*, 2007, 38, 673-683.

[34] Kim, U. J.; Gutierrez, H. R.; Gupta, A. K.; Eklund, P. C. *Carbon,* 2008, 46, 729-740.

[35] Kim, U. J.; Furtado, C. A.; Liu, X.; Chen, G.; Eklund, P. C. *J. Am. Chem. Soc.*, 2005, 127, 15437-15445.

[36] Telling, R. H.; Ewels, C. P.; El-Barbary, A. A.; Heggie, M. I. *Nat. Mater.*, 2003, 2, 333-337.

[37] Lide, D. R. Handbook of Chemistry and Physics; *CRC Press*, 2003-2004.

[38] Hirsch, A. *Angew. Chem. Int. Ed.*, 2002, 41, 1853-1859.

[39] Ryabenko, A. G.; Kiryukhin, D. P.; Kichigina, G. A.; Kiselev, N. A.; Zhigalina, O. M.; Zvereva, G. I. Krestinin, A. V. *Dokl. Phys. Chem.*, 2006, 409, 181-185.

[40] Fujigaya, T.; Nakashima, N. *Polym. J.,* 2008, 40, 577-589.

INDEX

D

E

F